Introduction to the Constraints-Led Approach

Introduction to the Constraints-Led Approach
Application in Football

Ben Bell, BSc

Sport and Exercise
University of Gloucestershire
Cheltenham, England

ELSEVIER

ACADEMIC PRESS
An imprint of Elsevier

Academic Press is an imprint of Elsevier
125 London Wall, London EC2Y 5AS, United Kingdom
525 B Street, Suite 1650, San Diego, CA 92101, United States
50 Hampshire Street, 5th Floor, Cambridge, MA 02139, United States
The Boulevard, Langford Lane, Kidlington, Oxford OX5 1GB, United Kingdom

Notices
Knowledge and best practice in this field are constantly changing. As new research and experience broaden our understanding, changes in research methods, professional practices, or medical treatment may become necessary.

Practitioners and researchers must always rely on their own experience and knowledge in evaluating and using any information, methods, compounds, or experiments described herein. In using such information or methods they should be mindful of their own safety and the safety of others, including parties for whom they have a professional responsibility.

To the fullest extent of the law, neither the Publisher nor the authors, contributors, or editors, assume any liability for any injury and/or damage to persons or property as a matter of products liability, negligence or otherwise, or from any use or operation of any methods, products, instructions, or ideas contained in the material herein.

Library of Congress Cataloging-in-Publication Data
A catalog record for this book is available from the Library of Congress

British Library Cataloguing-in-Publication Data
A catalogue record for this book is available from the British Library

ISBN: 978-0-323-85026-1

For information on all Academic Press publications visit our website at
https://www.elsevier.com/books-and-journals

Publisher: Stacy Masucci
Acquisitions Editor: Elizabeth Brown
Editorial Project Manager: Pat Gonzalez
Production Project Manager: Kiruthika Govindaraju
Cover designer: Mark Rogers

Typeset by TNQ Technologies

Contents

About the author

Ben Bell is a full-time student, studying for an MSc by research in Sports Leadership, Education and Society with the University of Gloucestershire. Ben's main research area is coaches' behavior in football environments, applying social theory to sporting contexts. He is particularly interested in the work of social theorists such as Erving Goffman and Michele Foucault and also has an enthusiasm for the constraints-led approach after learning about it at undergraduate level, receiving a first-class hours degree in BSc Sports Coaching. Ben started his coaching journey delivering afterschool clubs working for a sport development company in the Vale of Glamorgan. He has coached grassroots teams in England and Wales while completing football coaching qualifications. Ben is now coaching at Penarth Town helping develop young people and has attained a UEFA B license with the Football Association of Wales.

Preface

The aim of this book is to connect theory with practice. To bridge the theory—practice gap, this coaching resource is made up of two parts: theoretical chapters and football/soccer practice designs. The Constraints-led approach (CLA) is a framework to aid the coaches to design their practice environment and not a method of coaching that dictates how practices should be designed. "The aim of the CLA is to identify the nature of interacting constraints that influence skill acquisition in learners" (Davids, 2011, P.3).

This book is a coaching resource and not a textbook; it is designed to give you an understanding of the CLA framework, thus allowing you to understand what tools can be drawn on in practical environments. The book is aimed at football/soccer coaches working with any age groups and any ability, aiming to deliver theoretically informed practices. Despite the book using football examples to illustrate points and including 40 football practices, the theoretical chapters also hold value to any practitioners looking for an introduction the CLA.

This book draws on literature from key texts such as Gibson, 1979 titled *The ecological approach to visual perception* and the book by Davids, Button, and Bennet published in 2008 titled *The dynamics of skill acquisition: A constraints-led approach.* These books provided a comprehensive theoretical overview of ecological dynamics and motor learning, respectively, from the perspective of the CLA. This book seeks to combine the existing literature to offer an introduction into the complex theoretical makeup of the CLA framework.

This book is also academically underpinned with content often paraphrasing existing literature. Some of the terminology and concepts may leave you feeling overwhelmed, but I feel as an author that it is my duty to include them. I feel I would be doing you a disservice to distort the content to an oversimplified version, even if it would be easier to understand because what you would be learning would not be a true reflection of the theory.

The theory section of the book is divided into three chapters: ecological dynamics, perception—action coupling, and the role of constraints. Giving you an understanding of the theory in a comprehensible way and allowing you to gain a deeper understanding of the theory. This is followed by a chapter called practice design overview, which provides the link between the theoretical section and the practice designs.

The ecological dynamics chapter provides an insight into a nonlinear approach to learning and discussing how constraints create affordances that learners react to.

The second chapter is focused on perception—action and how it exists in a continuous loop and is essential for learning to occur, outlining how we perceive affordances and how behaviors emerge using perception—action.

The third theoretical chapter is designed to explain the role that constraints have in the learning environment. This chapter builds on the previous two discussing the manipulation of constraints to create affordances in the environment. This is the view that learning occurs when constraints are imposed.

The fourth chapter is titled practice designs overview and is a link between the theory of the CLA framework explaining the unconventional template used for the practices and how this links to the theory. The practices are divided into three categories according to their focus: in possession, transitions, and out of possession. I encourage you to make the practices your own as they are easily adaptable to suit any environment with different numbers of performers and different ability levels.

This book is not a *how to use* guide for the CLA; it is an introduction to the theory and how it can be successfully applied in practice to benefit the learner/performer. This is certainly not the only way the framework could be used, but it can provide a clear picture of what practice environments can look like. This is because the CLA can be applied to any practice. The framework infers that particular practices are more effective for learning. For example, practices that are representative, include affordances, and force problem-solving better allow learning to occur. It is a framework that can be applied successfully and unsuccessfully. This is not to say that all the practices in this book are a collection of the most successful ways in which the framework can be applied. The practices designed in this resource are valuable as they provide the opportunity to see directly how the CLA can transfer into the practical environment. These are practices that can be implemented with an understanding of the theory underpinning them. A deeper understanding of the theory allows the coach to understand the value a practice can have for the learner.

This resource refers to "the coach" because that is exactly the role they must perform and they must coach. The CLA requires the coach to be actively involved in the environment as the coach is the architect of the environment. Referring to coaches or practitioners as learning facilitators can be misleading when the CLA is a framework that advocates for a hands-on approach. Crucial to the CLA is appreciating the mutual dependence the learner has with their environment (Gibson, 1979). The CLA framework requires the coach to be heavily involved in the learning process; this perspective challenges the perception of the game being the teacher. The skill of the coach comes in designing the practice environments, knowing what constraints to manipulate, when to increase or decrease the levels of complexity, and how to intervene either implicitly or explicitly.

The coach provides the intention for the practice from a full 11v11 game to a 1v1. Both 11v11 and 1v1 constitute a CLA and support the coach to better understand what constraints to manipulate to help support the learner. Pinder et al. (2009) tells us that affordances are invitations for action. Putting novice learners into an 11v11 or 8v8 or 6v6 depending on the age could easily overwhelm them as they are unable to react to the affordances in what is a very chaotic environment. This is due to it being beyond their control and due to their level of ability they are unable to engage with their environment, meaning it is too difficult for them to process all the information from what is going on around them.

When wondering whether *the CLA does actually work*, it is impossible to answer as the CLA is a framework and thus impossible to measure as one single coaching method. This is due to the framework being applied in many different practices and contexts. The CLA is a framework to help us understand behavior rather than a coaching method, which only includes certain types of practices such as small-sided games.

However, it is possible to see that learning occurs and there are ways to measure performance to see the outcomes. Deciding what to measure or look for is determined by the focus of the practice. As many coaches do not work at the elite level with access to a performance analyst, biomechanist, psychologist, and other support staff, it can be very difficult to collect quantitative data. It is still possible for performance to be framed around what success would look like in each practice. This requires the coach to be aware of the aim and use their observational skills. A practice could be designed to afford lots of crossing opportunities. The coach will have an idea if the practice is working if the performers are attempting many different crosses from different positions. In addition, notational analysis could be used to record the number of crosses and if they were accurate at finding a teammate or not. It is also worth taking into considerations the feelings of the performer to find out if they feel more comfortable performing a skill or playing in a chaotic environment or they just feel that their intensity has improved.

Acknowledgments

I acknowledge this is all my own work and can be used as a reference for future work.

Thanks to my friends and family who have supported me as I wrote this book.

Thanks to the university lecturers who taught me about the constraints-led approach.

Thanks to Bob Mckenzie who gave me my first coaching opportunity at Charlton Rovers and wrongfully trusted me so much.

Thanks to Fraser Rankin for letting me help coach his Churchdown team.

Thanks to Shaun Watkins and Pete Ingram for giving me so much freedom to experiment with new ideas, coaching the excellent boys at Penarth Town.

Most importantly thanks to the people I have coached.

The theoretical foundation of the constraints-led approach

1

The constraints-led approach (CLA) is a framework within which behavior can be viewed, explained, and changed. The underpinning of the CLA is ecological dynamics introduced in Gibson's book *The Ecological Approach to Visual Perception* (1979) which says that we, as organisms, are capable of problem-solving in our surroundings. The CLA in practice is the manipulation of constraints within the environment inviting learners to self-organize (Newell, 1986). It is important for football and sport generally, as it views the relationship between the athlete and the environment as critical for understanding performance and learning, rather than emphasizing personal qualities and ability.

Drawing upon research from ecological psychology, dynamic systems theory, movements structures, and chaos theory, the CLA framework has been used to understand how children and adults acquire and adapt movement skills (Davids et al., 2008; Davids, 2011). The coaching framework that was discovered following the development of understanding of ecological dynamics helps us understand learning.

Gibson (1979) discussed ecological psychology, and how the way an organism perceives an environment is the key to how the organism acts within it. Within this book, affordances are defined as opportunities for action. Affordances are available to organisms within the environment. For example, if we (as humans) saw a mouse, the reaction would be very different to the reaction of a cat, who could try to chase it and play with it and probably kill it. In this example, the mouse is the affordance as it is inviting an action from an organism in the environment. This shows how different perceptions lead to different actions as the way we perceive a mouse would be different to the way a cat does.

Ecological dynamics is made up of Ecological psychology and dynamical systems, addressing how nonlinear systems interact, as organisms ourselves we are all nonlinear systems. Combine ecological psychology and dynamical systems, and you have ecological dynamics, which underpins the CLA. Ecological dynamics tells us that organisms are nonlinear. Nonlinear systems do not develop in a step-by-step process with consistent gradual change; they instead stop, jump, and even regress in ways which can be difficult to predict.

The key thing related to practice is that the way we behave is dependent on what we perceive in our environment. The coach is viewed as an environment architect as they can control some but not all of what the learners perceive in the environment.

Introduction to the Constraints-Led Approach. https://doi.org/10.1016/B978-0-323-85026-1.00005-4

There will always be factors such as the playing conditions which the coach has no control over.

Building on this, what the learner/performer perceives can be manipulated by constraints.

Constraints on perception are divided into three: organism referring to the individual; task referring to the rules, size of the field, duration, number of goals, and number of players; and environment referring to the weather conditions and playing surface (Newell, 1986). The constraints that can be imposed on the organism, task, and environment are discussed in more detail in Chapter 5, Practice design overview, on pages 23—27.

Aspects of these can be manipulated to create emergent behaviors implicitly where organisms produce actions as a response to what they perceive in their environment. Out of the three, this book mainly focuses on manipulating the task and environment.

The CLA says that learning is nonlinear and following the introduction of innovative pedagogies, contemporary coaching is becoming more nonlinear (Chow et al., 2016). This challenges traditional coaching methods that argue that learning is a linear process which is why there has been resistance to innovative approaches to coaching (Butler, 2005). A CLA is important as it is a framework that can be applied to many sports (Magius et al., 2015; Renshaw et al., 2015). Despite the fact it could develop approaches to learning and skill acquisition, the lack of empirical research into the practical application along with the complex nature of ecological dynamics and perception—action could act as a barrier to it being successfully implemented on a large scale (Martindale & Nash, 2013; Renshaw et al., 2015).

Crucial elements of the practical application of the CLA are these four principles. (Renshaw et al., 2019).

1. Representative design
2. Affordances
3. Intentionality
4. Repetition without repetition

Representative design refers to how realistic a practice is ensuring what the learners are perceiving in the practice environment reflects what they could experience in the performance environment. Affordances refers to what is in the environment for learners to react to in accordance with the practice intention. Intentionality is the aim of the practice acting as an overarching constraint of behavior. Finally, repetition without repetition refers to the amount and quality of different possible solutions to problems created in the environment.

These principles allow the coach to understand the value of a practice and can be seen as dials that can be adjusted. On a speaker, there are dials that control bass, treble, volume, echo, and tone. In coaching, these four principles can be adjusted like dials on a speaker. For example, a coach could dial up the affordances in an environment while dialing down the representative design. It is a way to help inform practices and understand why behaviors emerge within them.

Ecological dynamics

The aim of this chapter is to explain ecological dynamics and how it underpins the constraints-led approach (CLA). Ecology refers to environment, and dynamics is a large and diverse set of methods and concepts. Ecological dynamics suggests that humans are self-organizing dynamical systems, and our behaviors are reactions to the environment. Self-organizing refers to our ability to problem-solve. This is based on ecological psychology being the study of information transactions involving a living system and their environment (Davids et al., 2008).

Ecological dynamics can be applied to football as it considers performers and teams as adaptive systems that are a complex network of integrated subcomponents that interact, such as each performer in a team (Renshaw et al., 2019). An ecological approach puts stress on the individual and the environment and uses affordances as opportunities for action (Araujo et al., 2006, 2010). This means that our behavior is a result of how we react to things in our environment. Dynamical systems theory is an alternative paradigm to traditional coaching methods and suggests that movement control is not centrally stored (Barris et al., 2013), signifying its link to self-organization, coordinative structures, and perception—action coupling, which is discussed in more depth in the next chapter.

Gibson (1979) tells us that performers make up their own environment. Without the performers, the environment is not the same, showing that the organism and environment are mutually dependent on one another. However, the environment can become complex because there is a crossover where environments of performers merge. This can occur when a performer can be perceiving at the same time as affording opportunities for other performers to act. Furthermore, Gibson (1979) stated that performers are the most complex objects of perception in the environment meaning we can react to what happens in the environment.

As mentioned, ecological dynamics suggests we learn by self-organizing as we are dynamic systems capable of finding our own movement solutions. The CLA characterizes performers as nonlinear dynamical movement systems, and this provides a basis for nonlinear pedagogy (Chow et al., 2006; Davids et al., 2008). The CLA accepts people find their own movement solutions, and if the outcome is successful, there is no need to interfere with someone's movement pattern. However, these solutions will not look the same for everyone due to physical constraints such as height and weight (Araujo et al., 2006).

The CLA advocates that learning is a nonlinear process, whereas a traditional approach to football coaching is far more linear, involving learners trying to

Introduction to the Constraints-Led Approach. https://doi.org/10.1016/B978-0-323-85026-1.00008-X

replicate a perfect model of a skill. Learners will do this until it is performed correctly with the view that through repetition, they will be able to do it automatically and then focus on decision-making.

An ecological approach encourages decision-making to always be involved in the learning process. Despite performers finding their own solutions, the coach must be actively involved in the learning process, as they have an input in the design of the environment in which the performers self-organize. When designing a practice, the coach should create a problem in the environment that the performers solve, and the solution to the problem is linked to the aim of the session. This is the principle of intentionality, which is an overarching constraint in the environment.

Ecological dynamics can help to inform decision-making using intentionality. Intentionality refers to the aim for the learner, for example, a baby climbing upstairs toward a parent/guardian. Their complex neurobiological system of a baby is composed of many degrees of freedom. As the intention of the baby is to get to the parent/guardian, the baby self-organizes to find a movement solution (Davids et al., 2008). In this example, the stairs are the affordance that a baby reacts to, as the stairs invite the baby to crawl.

The underpinning of the CLA in practice is the manipulation of constraints to create affordances within the environment inviting participants to self-organize (Newell, 1986).

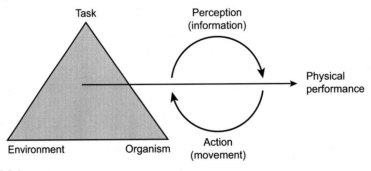

FIGURE 2.1

Newell's (1986) model of interacting constraints.

This model demonstrates how the CLA works. It is comprised of two features, but both are connected and lead to behavioral change. The first part is the triangle that shows constraints can be applied to manipulate the task, organism, and environment. The second feature shows that within the environment, participants self-organize using perception−action and this leads to physical performance that refers to the behavior of the performer. Perception−action is how the information we already have informs our actions.

This model is excellent, but future research could build upon Newell's model (1986) to include the notion of "culture" within an environment. The framework advocates that intentionality shapes decision-making and how important the environment is in shaping intentionality (Araujo et al., 2006, 2010). To take a football example, it is common for coaches to have a playing philosophy, and the culture they are involved in creating could impact on a player's decision-making. If the head coach has a philosophy that involves players being brave in possession, this could impact decision-making, perhaps leading to players dribbling in their defensive third or playing more high-risk passes than they would normally.

Degrees of freedom

Referring to Fig. 2.1 Newell's model of interacting constraints (1986), the organism refers to the individual/learner/performer. Human beings are coordinated structures with many degrees of freedom. The degree of freedom is the number of ways a system can move; for movements to be performed, multiple structures need to be controlled and coordinated (Pinder, 2011). This means that we are made up of many degrees of freedom, and it is controlling these degrees of freedom, which allows us to move.

Degrees of freedom help show the huge number of variables that are in play when moving, joints, nerves, muscles, and bones. The number of degrees of freedom grows exponentially when considering everything in play with a movement.

Many movements require degrees of freedom to be controlled to coordinate movement outcomes (Newell & Ranganathan, 2011). Movement outcomes are a result of using perception—action to self-organize in the environment.

To move effectively, these degrees of freedom need to coordinate. Even systems that are not necessary for a movement need to be controlled for a movement to be successfully produced. For example, when shooting with their stronger foot, performers will most likely use their arms to balance and the trunk will rotate as the striking foot goes back. But when shooting with a weaker foot, performers may sometimes forget to move their arms or their trunk does not rotate as much. This leads to it appearing and feeling different. This is likely due to degrees of freedom being frozen as they try to coordinate the movement. But as performers practice, they become better at coordinating their movements as they have better control of the degrees of freedom.

Level of complexity

Moving on, the level of complexity is vital for learning. The word "complex" derives from the Latin *complexus* defined as interwoven describing a network of interacting components with each component having the ability to interact with another (Davids et al., 2008). The theory of complex systems tells us that order will emerge in chaos under constraints. Thus, if a practice is too complex, learners will not be able to problem solve because it is beyond their capabilities (Renshaw et al., 2010).

The level of complexity must be designed so the outcomes are reachable for the learner; this is similar to Vygotsky's (1978) zone of proximal development (Daniels, 2005). This shows how the level of complexity comes into ecological dynamics as performers need to self-organize for learning to occur; if the environment is too complex, they will not be able to self-organize.

We know from ecological dynamics that organisms self-organize. The difficulty of the problem must be aligned to the skill level of the learner. As order will emerge in chaos under constraints, for football there should be some notion of chaos, and it is through the chaos where learning can take place. Although the level of chaos will need to suit the capabilities of the learners and be working toward a directed goal, this will allow learners to problem-solve effectively (Renshaw et al., 2019). For example, an 11v11 game would be too complex for a novice, whereas a 3v3 would be more appropriate, as there are less affordances to attune to in the environment, meaning there is less going on around the novice performer making it easier for them to learn in that environment.

An example of where coaches try to reduce the complexity is with a small sided game (SSG). An SSG is an example of the task being manipulated by reducing the number of players, goal sizes, goal locations, and pitch dimensions; further constraints can also be imposed within an SSG (Hill-Haas et al., 2011).

In recent years, there has been a discourse created claiming that the game is the teacher. The CLA is a framework where the coach is actively involved in the learning process. Issues arise as according to the CLA, a 5-a-side game, for example, would not be the most representative environment for performers as the affordances in the environment are not representative, mainly due to the pitch dimensions and where the goals are located. Therefore, performers will be producing movement solutions that are not representative of the performance environment (Sampaio et al., 2014). Referring to the four principles as dials, representative design would be low as would affordances.

However, the intention of practice is key, as a critical assessment of the CLA could argue that SSG has been shown to increase physiological capacity due to the repeated short sprints required as well as improving technical attributes. The manipulated practice influences emergent behavioral dynamics. Coaches could design this into practices such as high-intensity interval training to increase fitness levels. Increasing fitness levels improves the ability of the performer to make decisions as fatigue can have a negative impact in the way performers make decisions (Sampaio et al., 2014). SSG's have been shown to increase levels of acceleration, power, speed, and speed endurance (Stevens et al., 2015). However, one reason representative design would be low is that in an 11v11, players will have to run 50−70 m continuously in matches and might not have been exposed to those distances in an SSG due to the area size.

On the other hand, the task constraints invite players to control the ball in tight spaces and coadapt with each other (Hill-Haas et al., 2011). Coadaptation is "the continuous interactions that emerge as athletes co-adapt to each other's behaviors while self-organizing" (Renshaw et al., 2019, p. 110). This means that due to the

5-a-side game having different goals, different numbers of players and a different pitch dimensions the performers would behave differently in a 5-a-side game than they would in an 11v11 match.

This shows how different practices can be understood using the CLA framework, as well as the fact that having a balance when training is key. 11v11, 8v8, or 6v6 depending on the age is as representative as can be but does not offer many affordances. In addition, only using SSG would not be effective as the performers will not have been exposed to enough problems that they will realistically have to solve in the performance environment. With an SSG, the task has been manipulated, thus offering different affordances. Therefore, an SSG could be used as a tool for performers looking to build up fitness and technical attributes.

Skill acquisition and adaptation

Moving on to skill acquisition and adaptation, traditionally, motor skill acquisition has been researched as an internal process that brings changes in the learner's movement capabilities. Skill acquisition links to the CLA as it involves the learners interacting with the environment, detecting information, and appropriately timing movement responses (Davids, 2011a, 2011b; Renshaw, 2011).

This should result in coordination patterns that adapt to a range of performance settings. This means performers react to what is in their environment and how they behave is a result of that.

Skill acquisition is the initial process of learning something, and skill adaptation is being able to apply it in different scenarios with different pressures (Davids et al., 2008). CLA is an innovative approach to learning focused on skill acquisition and adaptation; it suggests that humans can find their own movement solutions and learn through perception–action. For example, learning to pass the ball is acquiring a skill, whereas applying those passing skills in a chaotic performance environment under pressure such as a match requires performers to adapt the skill.

Skill acquisition is separate from skill adaptation, as learning is a gradual process and should result in behavior which is less susceptible to factors such as fatigue and nerves. Learners must acquire skills and then learn to adapt them to the environment. A key aspect is the adaptation of the skills as the task and environment can change (Davids et al., 2008; Newell and Ranganathan 2011). To be able to pass in that environment requires them to adapt their skills. Being able to do it in different contexts with variability requires skill adaptation. This framework suggests the best way to develop skill adaptation is to practice in a realistic environment.

Introductory research on skill acquisition used Newell's (1985) framework; this stated three stages of learning:

(1) Coordination, focusing on the assembly and functionality of coordination patterns.
(2) Control, greater attunement to dynamic performance environments.

(3) Skill, exploiting intrinsic dynamics and being more flexible to exploit environmental sources of information (Newell, 1985, Cited in Davids et al., 2008).

Newell himself identified the flaws in this model as only being applicable to serial tasks such as hitting a ball and not continuous tasks where strategy and decision-making hold more significance than an individual's coordination (Renshaw, 2011).

These three stages of learning can struggle to be applied to football as many different skills make up a performers' overall ability. Performers can be at stage 1 of learning in one aspect but stage 2 at another. This could lead coaches to focus on the coordination phase, with coaches designing practices that break down learned complex movement patterns in an attempt to remodel coordination patterns. This can have significant consequences on the psychological state of an individual impacting their confidence and subsequently their capabilities (Davids et al., 2008). This is more common when using a traditional coaching method where learners are tasked with repeating movements trying to replicate a perfect model.

Therefore, this model using the three stages of learning can be used when trying to learn individual skills such as ball mastery work or during 1-2-1 coaching sessions where the learner is trying to complete tasks such as shooting. However, it does not transfer successfully to more chaotic situations where a learner needs to make decisions based on what they perceive in their environment. The three stages of learning could be applied to an individual's ability to pass and receive the ball comfortably. However, it cannot be applied to the movement required to create the space to receive or the decision-making of what type of pass, where to pass, and who to pass to.

To summarize, ecological dynamics underpins the CLA framework. It is comprised of two parts: ecological psychology and dynamic systems theory. It suggests that learning is nonlinear and that learners self-organize against affordances in their environment using perception—action. Furthermore, to be able to self-organize, the level of complexity needs to be appropriate for the learner; if the environment is too complex, the learner will not be able to attune to all the affordances in the environment.

Chapter key points

- Learning is a nonlinear process.
- Humans are organisms who learn by self-organizing against affordances in their environment.
- The environment must be the correct level of complexity for learning to occur.
- Intentionality informs decision-making.
- Skill acquisition is the initial process of learning.
- Skill adaptation is application of what has been learned.

Perception–action coupling

3

The aim of this chapter is to build on what has been discussed previously about ecological dynamics, complexity theory, and skill acquisition and adaptation. Referring to Figure 2.1 Newell's model (1986) on page 4 in Chapter 2, this chapter focuses on the second part which is perception–action, and these two exist in a continuous loop; hence they are labeled as coupled.

Crucially, Gibson (1979) highlights that movement generates information which goes on to support further movements. Perception refers to information already in the system, and action refers to movement. Information can be gathered from anything that is in the environment such as players or location of the ball (Vuuren-Cassar et al., 2014). Pinder et al. (2009) emphasizes how experts can perceive information more efficiently than amateurs. The CLA framework is underpinned by ecological dynamics, and perception–action comes into ecological psychology.

It is vital to understand that perception–action is a constant loop. What we perceive is taken in by the brain, which leads to an action to be performed. This action leads us to perceive further in a new environment. Moving around in an environment can directly impact what is perceived and thus how we act. This shows how movement generates information, which goes on to support further movement.

Interdependency between the generation of movement and the perception of information has grown within neurobiological systems and led to them being studied together (Martindale & Nash, 2013). This advocates that information and movement have a cyclical relationship, and for this approach to be effectively used, practices must be designed with a tight coupling of perception–action (Davids et al., 2008; Pinder et al., 2009). This is because performers attune to information in the environment, thus producing movement as a response as information and movement are interwoven (Pinder, 2011).

It is important to understand how the theory of perception–action coupling can be applied. Traditional approaches to coaching football have often focused on technique and the ability to execute a skill close to a perfect model (Hargreaves & Bate, 1990). Due to the discourse created by this approach, many practices ignore decision-making; as a result, perception–action has been decoupled due to the lack of external instability (Dicks & Chow, 2011). This is a problem with unopposed practices.

The CLA suggests learning cannot take place without perception–action, as performers will not understand decision-making in a match as the skills acquired in the practice environment do not transfer to the performance environment (Renshaw et al., 2019).

This is significant as it shows that perception—action is used to improve performers' abilities to make decisions (Pinder et al., 2009). Perception—action coupling can be determined as effective, if the performer can achieve a goal accurately, and efficiently, it is likely to increase the level of performance. This is the crucial point to consider for coaches as there should be a transfer from the practice environment to the performance environment (Renshaw et al., 2019).

This is important for coaches as practices should be designed with a tight coupling of perception—action which involves providing information in the practice environment that would be present in the performance environment. The information is critical as it is the information that, the performer perceives, informs what movement they produce. This is how we can see behaviors emerging using a CLA framework. In addition, this will better allow performers to feel more comfortable in the performance environment as they would be used to reacting in similar environments when training (Araujo et al., 2006). By developing performers to be more comfortable in the performance environment, this could lead to an increase in performance as they would have learnt more in practice.

Practices should offer the performer enough relevant information to produce movement solutions which can transfer to the performance environment. Unopposed practices do not provide performers with the relevant information, thus decoupling perception—action. Which means performers are unable to learn as they cannot self-organize as there is no opposition to react against, to build up perception-action synergies. For example, an unopposed passing drill where the coach demands players scan (look around) before receiving a pass decouples perception—action. This is due to there not being other performers around in the environment to scan for. Contrast that to a performer scanning their surroundings and moving into space in a match. The performer scanned to find space (perception) and moved into it as a result (action). While moving into space and when they arrive in space, they are still perceiving as they are moving. This shows how perception—action comes into ecological dynamics as movement generates information, which goes on to support further movement. Perception—action is crucial for decision-making, and the CLA says that these two must always be connected for learning to occur.

Another football example would be a midfielder seeing that their winger has possession in the final third. The midfielder thinks the winger might cross so they decide they want to get into the box.

This contradicts research into linear behaviorist coaching methods focused on technique, which advocates for high amounts of repetitions of an individual task (Hargreaves & Bate, 1990). There are other nonlinear approaches to coaching, but research around TGfU and Game sense does not address perception—action although within the approaches there would be some overlap as practices would appear similar (Renshaw et al., 2015).

The performer then must assess other performers in the environment both opposition defenders and teammates. The performer then must decide the best route to the penalty box and whether to run, sprint, jog, or walk and the appropriate direction. As they begin running, a gap appears between two defenders, which invites the performer to speed up to run into the gap, so they increase their speed to get on the end of the cross. The speed of the performer's initial run gave them the opportunity to perceive the gap, which informed the subsequent action. This shows how perception leads to movement, and this cycle continues showing how perception–action is coupled. With an unopposed practice, the defenders would not be there, and thus, perception–action would not occur; we know the CLA advocates that for learning to occur, perception–action must be coupled.

The ability to execute skills at correct times in football requires accurate perception of information coupled with movement coordination. Skilled perceptual anticipation is crucial for successful physical performance and has brought sport science to research perception–action (Pinder et al., 2009). Information can be gathered in the system using the senses; in a 1v1 situation, for example, players can perceive an opponent's movements. An attacker would try to move in a way to deceive the defender to get past and move forward (Araujo et al., 2006).

With every second, the attacker waits that the information is susceptible to change imposing new constraints encouraging new movement solutions. To improve this skill, an attacker should be put in similar scenarios during practices to problem-solve. This will provide opportunities to discover new movement solutions due to the repetition without repetition and develop the ability to interpret perceptual information earlier, thus improving decision-making. With practices that are designed to couple perception–action, there is lot of instability and variability in the environment that the performers react to. This highlights how the CLA can be used in football as perception–action is tightly coupled (Araujo et al., 2010; Barris et al., 2013). All the practices in this book are designed with a tight coupling of perception–action.

Part of the coaches' role is to manipulate the information in the environment which alters what the participants attune to, thus producing movement as information and movement are interwoven (Araujo et al., 2006). The ability to execute the right decision at the precise moment where it has the maximum impact requires accurate perception of information along with movement coordination (Dicks & Chow, 2011). During a match, there are many moving parts and many decisions to be made. For these decisions to be made efficiently, participants must have developed the ability to perceive information as it changes in the environment. It is these changes that afford multiple solutions and decisions to be made (Dicks & Chow, 2011).

Vaeyens et al. (2007) conducted a study looking into decision-making, a trend emerged that as the number of performers playing increased, the decision time became slower and response accuracy decreased. This suggests that it is harder to make decisions in more complex environments. Pertinent to this is that many sports are complex with moving parts, thus making perception–action critical underlying the importance of the CLA (Davids et al., 2015). This study also backs up the point made that for learners to self-organize, the level of complexity must be correct so they can make decisions based on the affordances in the environment.

A crucial element of a CLA is understanding how it works in practice. Renshaw et al. (2019) discuss four principles to a CLA in practice.

1. Representative design: How realistic is the practice compared with the performance environment.
2. Affordances: These are opportunities to act and make decisions.
3. Intentionality: The practice should be aligned to the aim of the session.
4. Repetition without repetition: Participants should be able to perform the skill required but in many different situations.

Each of these principles should be viewed as dials on a speaker that can be adjusted. In the same way to get a deeper sound, the bass can be adjusted; practices can be manipulated to alter what the performers do. This book uses a scale of 1—10 with 1 being as low as possible and 10 being as high as possible. An 11v11 match would be a 10 for representative design if coaching an adult team as it cannot be more realistic compared with the performance environment.

Repetition without repetition can appear complex at first. This comes from the idea that to improve, performers should not repeat the same action but repeat variations of solutions in different ways, forcing performers to have to continuously solve problems as they would have to in the performance environment. From an ecological perspective, repetition without repetition means continuously attuning to the affordances in the environment. It is important to note for this to be applied successfully in practice, the design must ensure that multiple solutions can be found by the performers. This forces performers to find different solutions due to affordances creating different problems.

Intentionality is an overarching constraint impacting decision-making. It refers to the aim of the practice. If the aim of the practice is to improve shooting, then the performers should be required to shoot in many different ways from different positions under different pressures. Within the practice, there should be problems that the performers have to solve. The aim of the practice should be the solutions to the problems. For shooting, a practice could look like an SSG where the goals are brought close together. This affords the performers more chances to shoot. This also gives the coach lots of opportunities to coach shooting. This practice is designed called shooting SSG on page 73.

Another key point of perception—action is what performers perceive is specified by permanent features in temporary states, such as height and preferred foot, the direction in which a player is moving, the speed of this movement, and visible signs of fatigue or emotions (Araujo et al., 2009). These permanent features in temporary states form interactions, both social and behavioral, which are dependent on the performers communicating in the environment. This can appear confusing, but it is simple in practice. A football example would be late in a match a midfielder playing a pass to the stronger foot of their winger. The permanent feature is their stronger foot, and the temporary feature is the fatigue from playing the match. Perhaps earlier in the match, the pass would be played in front of the winger to run onto, but due to the fatigue, the decision is made to play it to feet.

If an improvement in performance levels is related to improvements in interactions by perceiving other performers, not only is learning highly complex but also the presence of other performers in the environment is vital for learning to occur.

Rondos are commonly used in football coaching due to the rise of Guardiola's Barcelona team and the ease in which they transfer onto social media. Rondos are a very simple game consisting of attackers trying to keep the ball away from defenders, very similar to piggy in the middle.

A rondo can be used successfully if the aim is to find the free man. In a 4v2 rondo, with four attackers and two defenders, the performer in possession should have two options cut off by the two defending players leaving one teammate available to receive a pass.

This is an affordance created by the 4v2 providing the performer in possession with the opportunity to pass to the free player.

However, if the rondo is performed with the intention of improving the defenders, then it is not a successful way of using the framework of the CLA and is far less useful for learners, as there is no consequence for the defenders for defending poorly. If a pass gets played between two defenders in a game, this would often lead them to running back toward their own goal as the attackers progress. But in a rondo, they just turn around, and the previous situation is recreated instantly.

The same applies if the intention is to be better at playing forward or in fact progressing the ball in any direction. In an 11v11 match which is the performance environment, if a performer receives a pass, it is possible for the performer to receive on the half turn to play forward or progress with the ball. If a performer splits two defenders with a pass and their teammate receives the ball in between the lines, the last thing a coach wants is that player then pass it straight back to where it came from. Rondos only have value if the intention is to find the free player. For this reason, it is only included in this book as a warmup with the intention of finding the free player on page 37.

Performers can coadapt as they use perception—action to discover movement solutions in their environment. Coadaptation can be used to change behaviors of the participants; opposition provides unpredictable dynamic instability that players self-organize against while building perception—action synergies that are representative of the performance environment (Renshaw et al., 2019).

Overconstraining is when there are too many constraints applied to a practice that they interfere with participants' ability to self-organize (Newell & Ranganathan, 2011). This modifies behaviors of participants as they are unable to coadapt.

A well-designed practice will allow the participants to attune to the affordances in the environment (Davids et al., 2008). Some coaches use the ubiquitous "5 passes before shooting" constraint; this is a badly designed constraint as it invites movement solutions that are not representative of the performance environment.

This is because perception—action is decoupled, which interferes with the natural process of self-organization based on the perceived affordances in the system

(Pinder et al., 2009). On reflection, this is a poorly designed constraint as it invites movement solutions that are not representative of the game. The defending team know that the opposition cannot shoot until they have made enough passes, which interferes with the natural process of self-organization based on the perceived affordances in the system.

Applied research into penalty kicks

This subsection focuses on applied research looking into what performers perceive in their environment. It is possible to conduct studies assessing what is perceived in environments with only a few moving parts creating variability, such as a penalty kick.

The importance of the role perceptual skills have has led to a rise in research around the ability to anticipate based on what is perceived in the environment and promote skill acquisition and adaptation (Pinder, 2011).

The penalty kick is a crucial set piece for football, mainly involving a kick taker and a goalkeeper trying to deceive each other (Hunter et al., 2018). There are two prominent ways to kick a ball for a penalty, using the instep for power and the side of the foot for accuracy. Goalkeepers can search for cues during the run-up to anticipate where the kick will be directed and the power of the ball. The most reliable source of information is the standing foot (Hunter et al., 2018; Lees & Owens, 2011; Savelsbrough et al., 2011).

Research tells us there is a decrease of 85% of the maximal speed of the ball when a kick taker is aiming for a target in the goal (Andersen & DÖrge, 2009). Andersen & DÖrge, (2009) concluded that when players can select their run-up, the penalties have 50%—60% more power and accuracy. Hunter et al. (2018) mention that faster shots result in early dives from goalkeepers and were less likely to be saved. This research would suggest that the kick takers are more likely to be successful using their own run-up and aiming to get as much power as possible.

This research into the intra coadaptation of goalkeepers and kick takers during the run-up and kicking phase influences how coaches should deliver sessions.

From the research, we know that kick takers are more successful using their own run-up, followed by a powerful shot using their instep (Andersen & DÖrge, 2011). This suggests that for a kick taker, perception—action is not pertinent, and therefore, the CLA is not the best approach to analyze the kick taker. Alternatively, for goalkeepers, the focus should be on the kick taker's standing foot at the point of impact. From this use of visual anticipation, they can preempt power and direction of the kick and make decisions based on this (Lees & Owens, 2011).

This highlights the strength of the practical application of perception—action as visual cues can be utilized as information demonstrating that the ability to understand information and make decisions as a result is critical for goalkeepers. From this research, we know that goalkeepers search for visual cues around the standing foot to give an indication of direction and power. In addition, the findings from this research are slightly outdated. It is now common for penalty kick takers to watch the goalkeeper up untill the point of striking the ball as they search for visual cues from the keeper to help inform which side of the goal to shoot. These shots often have less power as the kick taker is aiming for a target in the goal and running up slowly as they try to force the goalkeeper to move before they shoot.

To summarize, organisms perceive affordances in their environment. Well-designed sessions allow performers to attune to information in the environment where intentionality is not prescribed (Davids, 2011a, 2011b). Coaches can manipulate constraints in the environment to create affordances (Tan et al., 2017). The key benefit of this framework is that performers develop skills and decision-making in realistic practice environments, so they transfer into the performance environment, thus increasing their level of performance (Renshaw et al., 2010). The strength of this framework is that perception—action develops participants' ability to react against information in the environment and make decisions (Pinder, 2011). It is strongly suggested that for learning to occur, perception—action needs to be tightly coupled (Chow et al., 2016).

Chapter key points

- Perception—action is a continuous loop.
- Movement generates information that supports further movement.
- Manipulated constraints create affordances.
- Learning occurs when we use perception—action to react against affordances.

Role of constraints

4

The constraints-led approach (CLA) suggests that learning occurs when constraints are imposed on the learner; these constraints can be directed upon the organism, the task, and the environment as shown in Figure 1, Newell's model (1986). Task refers to rules, organization, and pitch sizes; organisms refer to the individual, and the environment refers to factors such as the weather and surface type (Araujo et al., 2010; Tan et al., 2017). The CLA framework advocates that people learn by reacting to affordances in their environment. Davids et al. (2008) tell us that a constraint limits the motion of parts within a system.

Constraints are imposed to create affordances. As discussed in the earlier chapters, affordances are opportunities for action within the environment and should be seen as invitations to act. Affordances are everywhere; a chair can be an affordance as it gives you the opportunity to act; in most cases, the action would be sitting down. Adjusting something in the environment creates an extra affordance. For example, replacing a normal football with a futsal ball affords more opportunities to dribble as a futsal ball is smaller and heavier.

Gibson (1979) discussed how we do not think like computers, but we directly interact with our environment. Affordances are linked to what organisms perceive in their environment. Furthermore, organisms search for affordances to perceive within their environment. This is also known as attuning to the affordances. This impacts practice designs. If practitioners want to elicit a change in behavior, there needs to be constraints imposed creating affordances that the performers must attune to.

This is complex as it is possible for different organisms to find different affordances within the same environment or react differently to the same affordance (Gibson, 1979). This can be due to physical differences such as height or weight, for example. Imagine a large hurdle; a tall person will seek to jump over it, whereas a smaller person will seek to crawl beneath it. The hurdle, in this example, is the affordance and has invited an action, but the action is different for the two different organisms. Psychological factors such as emotions can also have an impact on the way we attune to the affordances in the environment.

Experiences can also help inform our decision-making impacting how we act. For example, a person in a sports hall finds a small but heavy ball used to practice shot put safely indoors. They pick it up with their hands and discover that it is very heavy and thus puts it back down on the floor. That is them using perception–action with the ball as the affordance and the action was picking it up, they then perceived the weight of the ball, and the weight of the ball is the constraint.

Introduction to the Constraints-Led Approach. https://doi.org/10.1016/B978-0-323-85026-1.00009-1

17

Later, another individual sees the small ball on the floor and runs toward it and tries to kick it, only to suffer a painful injury. They did not have the experience of the previous individual; therefore, their action was different. This shows how important it is to consider the affordances within the practice design.

The ecological perspective is based on assumptions about performers attuning to ecological constraints, which underpins successful decision-making in sport. Complexity theory emphasizes the view that decision-making is based on interactions that are highly context specific (Newell, 1986). If an environment is highly complex, it is normally because there are lots of variables within it. This means there are lots going on, making it difficult to make sense of it and make decisions.

An alternative criticism of the CLA is that it fails to acknowledge the psychological aspects involved with sport due to it being a social constructivist theory. This approach neglects cognition and an individual's capability to understand complex systems (Davids et al., 2008). It suggests that as humans are self-organizing, they will solve problems that are presented to them. This is a weakness of the approach as decision-making occurs between interactions in specific contexts and the information in each of the systems involved shapes what movement solutions are produced (Pinder et al., 2009).

Before a football match, some teams do a drill involving the coach as a target player and a player passes the ball and gets it back and shoots unopposed at the goalkeeper.

A CLA would say that this practice will not help performers learn (Renshaw et al., 2010, 2019). Keeping the intention of shooting a more successful practice for learning would be a 2v1 situation where the attackers have 4 seconds to get a shot away as that is more representative and more affordance driven and will include repetition without repetition and maintains the intentionality as an overarching constraint (Chow et al., 2016).

But the fact many elite teams use this drill along with other unopposed drills suggests that the reason could be that the players achieve success and therefore are more confident in the match, thus improving their decision-making as confidence can improve capability (Araujo et al., 2009; Burton & Raedeke, 2008).

The CLA emphasizes that the learning facilitator is the coach, and part of the coaches' role is to allow participants to find their own solutions to problems in the environment and not break down and remodel existing coordination patterns (Davids et al., 2008).

Some coaches also deploy the "two-touch" constraint to encourage fast passing and movements. The idea, being if performers can only take two touches, they must play the ball quickly as dribbling is not allowed. The issue is that this prescribes intentionality that in turn decouples perception–action meaning learning cannot take place (Savelsbergh et al., 2011). This results in players not having to make decisions about when to pass, dribble, or hold onto the ball.

The intention was prescribed so players will move quickly because they know teammates can only take two touches so they need to be in a position to receive a pass quickly. However, once the first touch is made, the problem has been solved for the defenders as the players' next touch must be a pass or shot. These are examples of constraining to constrain rather than constraining to afford; this stops the natural process of self-organization based on the perceived affordances in the environment (Araujo et al., 2006, 2010; Renshaw et al., 2019).

This is because the two-touch constraint does not create an affordance, which forces performers to problem-solve as they would in the performance environment. This is a good example of using constraints to constrain rather than using constraints to afford.

Constraining to constrain such as the example provided alters behaviors of participants as they are unable to coadapt. Coadaptation can be utilized to change behaviors of the participants. In addition, the reason why unopposed practices do not support the learner to learn is an opposition provides unpredictable dynamic instability, which players self-organize against while building perception—action synergies that are representative of the performance environment (Renshaw et al., 2019).

Prescribed patterns of play

The aim of this subsection is to address how ecological dynamics helps inform the coach how the CLA framework can be useful when coaching patterns of play. Patterns of play in this section refer to coordinated, prerehearsed movements when in possession, instructing the ball go from A to B to C. It is common for football coaches to try to get their team to perform patterns when they play. An example of a pattern is the goalkeeper playing it to a right back; when the right back receives the pass, the winger ahead of them moves toward them coming short, with the aim of dragging the opposition left back with them. This creates space in behind, and the right back in possession plays a long pass into the space for a striker to run onto.

When coaching patterns of play, the intention is often to get a team to work on specific movements and repeat them until they become habitual. Issues arise when this practice is unopposed as despite the players being able to interact and execute the pattern of play, they are not decision-makers capable of perceiving information in the environment (Tan et al., 2017). This picks up on what Davids et al. (2008) discussed about practices being designed with a tight coupling of perception—action as unopposed practices decouple perception—action; therefore, learning cannot occur.

The main issue with prescribed patterns of play is that they rely on accurately predicting the movements of the opposition. No matter the levels of performance analysis or the amount of money available to football clubs, no one can know the future. Patterns of play are common, especially at elite level when a more linear approach is taken. Patterns of play can be very useful as it reduces the amount of decisions players make. With access to performance analysis, teams can understand how an opposition team plays and thus can train patterns, which they predict could be successful in the upcoming match.

The alternative paradigm is players relying on their ability to use perception—action to inform their decision-making, constantly scanning while moving and making decisions based on what they perceive, for example, scanning to find space, moving into the space while still scanning, receiving a pass, and scanning further to decide what to do next, pass, shoot, cross, dribble, or run with the ball. This can be very overwhelming especially if players are low on confidence and/or their level of ability is not as a high as their opponents.

Patterns can be practiced in many ways. To coach patterns of play using a CLA, it can be more beneficial for learning to use the notion of "ideas" rather than patterns. This is due to them being practiced in apposed environments, as Renshaw et al. (2019) tell us that opposition provides instability in the environment which performers then react against using perception—action.

The main difference being patterns prescribe intentionality and thus decision-making, whereas ideas allow for more freedom to the performers. The movements may be similar, but the players are allowed to make decisions based on what is in their environment.

Using the previous example, the idea would be to draw the opposition left back forward to open up space. If this is coached as an idea, players are still given freedom to make their own decisions. Therefore, if the opposition left back did not mark the winger and instead held their position, it would be better for the right back in possession to play it into the winger who would be in space. A prescribed pattern of play would dictate that the ball be played long into the wide area regardless of what the opposition does. The notion of "ideas" rather than patterns allows for decisions to be made by the performers based on the movements of the opposition.

The specific in possession practices in this book all allow for the coach to deliver "ideas" without the need for prescribing intentionality. This allows for perception—action to be tightly coupled as performers attune to affordances created by constraints in their environment.

It is still possible to develop patterns of play, but it is not possible to know future movements of the opposition. To use another example, the coach might want the team to play through the center of midfield.

A traditional method of using prescribed patterns could involve organizing a practice with two midfielders. A pattern could be that when a center back has possession, the deepest lying midfielder (pivot) makes a run forward with the aim of taking their marker with them. While they move, a more advanced midfielder drops short into the now vacant space, and the center back plays a pass to them. If this pattern were performed with lots of repetitions, a pattern would most likely develop. If practiced unopposed, this would become almost robotic in nature. This pattern would be unsuccessful if the opposition chose not to follow the midfielder moving forward and the space would not be created. Davids (2011a, 2011b) discusses the issue of fidelity as the movement patterns produced in the unopposed practices will not always be applicable to the performance environment due to the lack of affordances in the practice environment.

Furthermore, by practicing in an opposed environment, the coach could not only suggest ways for performers to create space using movement but also help them recognize when to stay still in space. Using the CLA framework, a more successful practice involves a 2v2 in a square which affords the performers chances to find and create space for each other without needing prescribed patterns. This specific practice is designed on page 53 called playing through central midfield specific.

Building on what Gibson (1979) tells us about organisms self-organizing, it would be better for learning if practices acknowledged that performers are not robots and can think for themselves. There should be opposition to make it more realistic by adding dynamic instability into the environment rather than demanding the players move in a prescribed fashion.

A key thing regarding the practical application of the CLA in football is coadaptation. This refers to the interactions that emerge as participants self-organize to achieve their goal (Renshaw et al., 2019). Another example is eye contact, when delivering a practice where the intention is for two strikers to combine in the final third, a constraint could be that the two performers must maintain eye contact as much as possible. This constraint creates an affordance, which invites the performers to interact and coadapt.

A key part of a team coadapting their movements is how successfully they interact with each other. Interactions will help the players to communicate and coadapt when playing to help the decision-making process. Interactions are ways of communicating. This involves players using body movements as one of many ways to communicate (Tan et al., 2017). For example, if a performer wants the ball to be passed to their right foot, they can signal by putting their right hand forward when calling for the ball. This helps the performer with the ball know exactly where their teammate wants the ball to be played. Alternatively, if a player is stood behind an opposition defender leaning forward, it might suggest that they want the pass played into space for them to run onto.

Practices do not have to be as structured as this to develop ideas. Coadaptation can be utilized in a team by the manipulation of constraints creating affordances. This can be done to manage an opposition by applying task constraints.

For example, if the coach wants to deliver a practice to get across an idea of how to play out from the back against a high press, the opposition can be managed using a task constraint such as a goal being worth three if they win the ball within 6 seconds of the goalkeeper taking a goal kick. This will afford the defending team to press quickly and aggressively making it potentially more realistic to the performance environment. This will also give the coach many opportunities to coach their ideas about how to beat a high press.

It is also important to note that if the coach only delivers practices getting across their ideas, this can also hinder the learning process as performers will become too reliant on the coaches' ideas. This can impact their ability to coadapt in the performance environment. As mentioned previously, a key skill of the coach is the ability to maintain balance with the types of practices. All types of practices have some value but being overreliant on any sort of practice, for example, SSG or specific practices to get across an idea as described in this section can negatively impact the performance level of the performers.

Interactions build on the idea posed by Gibson (1979) from ecological dynamics that humans self-organize in their environment. Interactions can help performers to communicate effectively in chaotic environments. As performers become more able to interact with each other, they will become better at coordinating their movements. When this is coupled with ideas, they have learnt in practice environments there should be an increase in performance levels without the need to rely on prescribed patterns. Instead, the reliance is on the performers to self-organize and use perception–action to coadapt as a reaction to the opposition.

Ecological dynamics can apply to football as within an environment coadaptation and interactions are important. If a team is able to successfully coadapt, they can have a better understanding of the movements of players on the pitch (Dicks & Chow, 2011).

Ecological dynamics underpins the CLA and says performers are equipped with the ability to problem-solve and make decisions using perception–action (Davids et al., 2008). For performers to be more successful, practices should incorporate the notion of coadaptation within ecological dynamics.

The practices designed in this book are all designed with a tight coupling of perception–action and constraints creating affordances that performers react against. Following the CLA framework, this creates a successful environment for learning.

Chapter key points

- Learning occurs when constraints are imposed on the organism.
- Learners can react differently to affordances in the environment.
- Learning does not occur in unopposed practices as perception–action is decoupled.
- Overconstraining can lead to performers being unable to coadapt.

Practice designs overview

The practices designed in this coaching resource are my interpretation of the constraints-led approach (CLA) framework and have been used in the settings I have coached in. These environments were all youth grassroots football teams in the United Kingdom. My intention is for the practices to be altered to suit the performers in your environment. This is because I believe sessions can have the most impact when they have been tailored to suit whoever is in the environment.

The practices are ordered: warm-up, in possession, transitions, and out of possession. In possession, practices have the intention of improving performers when their team has the ball. It can be focused on the performer with the ball, the team around the ball, and those away from the ball. Transition practices focus on the moment when possession changes over from one team to the other. Defensive transition is focused on what performers do when they lose possession, and attacking transition is focused on what performers do when they win possession. Out-of-possession practices have the intention of improving the performers whose team does not have the ball; again the focus can be on the performer on the ball and the rest of the team around the ball and away from the ball.

Practices are also labeled general, specific, or small sided game (SSG). General practices are less realistic than specific but are constrained to offer more affordances for skills to develop.

General sessions are opposed to provide realism and ensure performers get repetition without repetition. An example would be the practice called passing and turning general on page 45, where two teams play in a square and both have two goals to score in which are located opposite each other. Having two goals creates an affordance that performers can turn to attack a different goal frequently. This practice is not very realistic due to the multidirectional nature, but it provides more affordances.

Specific sessions are position specific with performers located in areas of the pitch in their positions. For example, a specific practice working on high pressing would be best performed in the area of the pitch where the high pressing would take place. This practice is called high pressing specific and is designed on page 105. Specific practices are more representative of the performance environment than general sessions but tend to offer fewer affordances.

SSGs are also used, offering more chaos than specific sessions. They can be less representative in terms of playing positions similar to a general session, which is due to performers being capable of self-organizing; therefore, the coach does not need to impose strict formations for an SSG. Instead constraints can be imposed to afford learning. Finally, an SSG will most likely afford the most fun to performers and is therefore a crucial aspect of any session.

Area size is important when coaching to ensure realism and intensity. The area size is not specified in the practice design. This is because the coach should be flexible and feel they can easily manipulate the area size creating different affordances.

The Football Association of Wales recommends a guide of 3 m^2 per person for an SSG for adults. Meaning a 5v5 would take place in a square 30×30 m. This can also be manipulated to offer different affordances. A smaller space affords more short passes; performers need good close control and will elicit more short sprints, compared with a larger space affording longer passing and longer runs (Hill-Hass et al., 2011).

The age of the performers and stage of their development will also have an impact. Smaller children, in general, will not be able to perform the same physical excursions as grown adults. In addition, if the coach wants to focus on improving endurance, larger areas would be best, whereas to improve speed and speed endurance, smaller pitches would be best. It is also important to consider intensity as larger area sizes can lead to a lower intensity.

In addition to these practices, coaches can plan their own practices using the CLA framework. Renshaw et al. (2019) discuss a step-by-step process that can be taken when designing a session.

- Step 1 is to have a clear intention. If the intention is to improve performers' ability to press high up the field, then the practice environment must be designed to allow high pressing to occur.
- Step 2 involves the manipulation of constraints. Using the same example, the manipulation of constraints should encourage performers to press high. It is important that the practice is realistic; the practice in this book called high pressing specific on page 105 begins with a striker, two wingers and a midfielder pressing a goalkeeper, back four, and midfielder who try to play out from the back into a mini goal.

  A task constraint imposed is that two more performers can help pressing after six seconds. This encourages the initial four to delay the opposition long enough to have support of another two. This offers more realism.
- Step 3 encourages the coach to not overconstrain as performers should be encouraged by the affordances created by the constraint to self-organize. This allows behavior patterns to emerge. This is because a key part is performers understanding when and how to use affordances.

Moving on, the layout of these practices is different to traditional session plans. Beneath a diagram of what the practice should look like, are the four principles that Renshaw et al. (2019) talk about and with a score out of 10. Think of these as dials that can be altered all the time. In adult football, an 11v11 match would be

10/10 for representative design as it is the most realistic environment compared with the performance environment on a match day. However, there are less affordances in an 11v11 match; there is less repetition without repetition, and the intentionality could also be weaker.

These dials are especially important as it allows you to understand how effective the practice could be. In some practices, the representative design has been dialed down to increase the affordances. This is where there is less realism, but there are more opportunities for the performers to practice what the coach is looking for.

Ecological dynamics tells us that humans will self-organize to find solutions to problems (Gibson, 1979). Therefore, each practice includes manipulated constraints creating a problem for the performers, and they must find the solution.

For example, a problem is that the goals have been turned around, and the only way to score is to receive a pass beyond the byline and then shoot. The solution is running off the ball because that is the only way to score. This gives the coach the opportunity to coach running off the ball and third man runs. This practice is called attacking third man runs SSG on page 75.

Instead of progressions and regressions, these practices utilize complexity theory, increasing or decreasing the levels of complexity. Traditional progressions make the practice harder often posing different challenges to the performers, and by doing that, the level of complexity is increased. Regressions make the practice simpler for the performers, and this is done by reducing the levels of complexity. There are many bullet points suggesting how to increase and decrease the level of complexity; these are not in an order and can be used freely and adjusted to suit the performers in your environment.

The constraint is what has been manipulated to create affordances in the environment. Many different constraints can be applied to create affordances, sometimes more than one and sometimes things the coach cannot control (Pinder *et al.*, 2009). For example, arriving at a session and discovering only half the space is available. The environment has been manipulated as the area size is half of what was planned for.

Like traditional session plans, many coaching points are listed. Some common themes reoccur in the coaching points such as creating and exploiting space, decision-making, and interactions between teammates. I would advise coaches to pick 2 or 3 and focus on them to avoid overloading performers with huge amounts of information. It is important to not become reliant on the coaching points; it is not a coach education course. It is more important to coach what you see rather than refusing to move away from your coaching points.

Referring to Newell's model of interacting constraints (1986) as shown in Figure 2.1 on page 4, there are 3 things that can be manipulated to form constraints: task, organism, and environment. It is impossible to know for certain everything that could be manipulated, but the following is a list including most of what could be manipulated by constraints.

Task

Competition: There should be winners and losers; this provides consequences if performers do something wrong. The coach should not be afraid to add competition for places if it suits the culture of the team.

Scoring systems: Alter scoring systems, goals can be worth more if performers do something aligned with the session intention, add rules where a goal only counts if performers produce movements that align to the session intention. The coach could add goals for creativity, positioning, disorganizing an opposition, etc. For example, to afford space to practice counterattacking, a constraint could be that goals only count if all players are in the opposition half coupled with counterattacking goals worth three. This practice is called counterattacking 6v6 SSG on page 89.

Pitch dimensions: Pitch can be altered to create different affordances: narrow/wide, short/long, half pitch/full pitch. For example, a wide pitch will afford players the chance to play the ball wide, whereas a long narrow pitch would afford more vertical passes.

Equipment: Equipment can be manipulated to create affordances. Coaches should experiment with different equipment such as futsal balls to encourage dribbling; additionally, if the balls are slightly flat, this will encourage dribbling. Taping half of the ball will offer different affordances by creating a slightly uneven roll. Different sized goals would also make a difference to where the performers aim shots and how the defenders behave.

Rules: The number of players per team, creating overloads and underloads. Timed constraints where more performers get added gradually to make the environment more chaotic.

Organism

Fitness: This varies based on the time of the season and the activity levels of the players outside of football. It also links to warm-ups and warm-downs and managing the workload during training.

Fatigue: This can be mental or physical and could be linked to the intensity of the session or duration of the practice. It can also be influenced by factors that the coach has no control over.

Emotions: This can be social bonds, psychological stresses, and different motivations players have can all be manipulated.

Playing style: Decision-making can be altered based on the style of play, e.g., possession or counterattack.

Confidence: Interactions with teammates and the coach as well as the perceived competence of the performer impacting their decision-making.

Environment

Surface type: This can be influential; most grassroots teams will play on park pitches, which can be muddy and uneven and have different gradients.

Weather: Weather impacts the organism and can be altered by deciding when and where to train. For example, training indoors in the afternoon will not be truly representative of an evening match outside. There are many factors, however, which often lead to training being held whenever is convenient, and that is fully understandable.

Club Culture: This links to the focus of the club or team. Is it a performance environment where results are paramount or is the focus solely on development?

Opposition: This refers to how the opposition plays, whether it is friendly, aggressive, or physical. This can be replicated in training by encouraging certain behaviors to make training representative of the performance environment.

Warm-ups

Dribbling at speed

Session created with SportsSessionPlanner.com

Representative design   3/10   Intentionality   8/10
Affordances   7/10   Repetition without repetition   5/10

Session organization: Four cones mark out a square with one player on each cone and another player in the middle. All players start with a ball. The aim of the players on a cone is to dribble to another cone. The aim of the middle player is to get to a vacated cone. All four players on a cone move at the same time, they cannot move diagonally. The four players on cones must work together to prevent the middle player from getting to a cone before them.

Problem: Can only move to a cone when it is free.

Solution: Interact with teammates to coordinate movements.

Increase levels of complexity:

- Add a 5 second timer when players are in the starting positions with individual players on a cone each and one in the middle waiting for them to move.
- Make the square of cones smaller.
- Adjust the ball players use to a smaller sized football or tennis ball.

Decrease levels of complexity:

- Make the square bigger and allow for longer rest periods after each round.
- Remove the footballs so the players are just focusing on interacting to coordinate movement patterns.

Constraint: Task is manipulated to afford moments where players must interact to coordinate movement patterns. Furthermore, with footballs, players not only have to recognize where to move, but they must also be skilled enough to dribble to a vacated cone.

Coaching points:

- Interactions between teammates.
- Maintaining control of the ball while moving at speed and quickly stopping and changing direction.
- Using both feet and different parts of the foot to manipulate the ball.

Dribbling chase game

Session created with SportsSessionPlanner.com

Representative design   4/10   Intentionality   8/10
Affordances   7/10   Repetition without repetition   6/10

Session organization: In a square, two players have a ball each. The aim of the players is try to tag each other in the square without losing control of their ball. One player starts as the tagger, and when they touch their opponent, the opponent becomes the tagger and the game continues to flow. If a player dribbles the ball out of the square trying to avoid being tagged, the player becomes the tagger. This can easily be adapted by having one big square or lots of squares set up next to each other.

Problem: Avoid or chase a player while keeping control of the ball.

Solution: Use changes of speed to exploit space, and dribble in different directions using the inside and outside of the foot; the tagger can try to get their opponent in a corner as there are fewer directions to move in.

Increase levels of complexity:

- Make the square smaller.
- Add more players into the square, adjusting the number of taggers to the preference of the coach.
- Use smaller footballs, tennis balls, or tape half a football.

Decrease levels of complexity:

- Make the square bigger.
- Adjust the ball by using sponge balls or large tennis balls.

Constraint: Task is manipulated by having two players with a ball in close proximity. The square forces the players to dribble in a confined space and the manipulation of the balls used will force players to react differently.

Coaching points:

- Using different parts of the foot to control the ball
- Creating and Exploiting space.
- Changes of speed and direction based on an opponents' movements.

Dribbling to keep the ball

Session created with SportsSessionPlanner.com

Representative design   5/10   Intentionality   8/10
Affordances   7/10   Repetition without repetition   5/10

<u>Session organization</u>: In a large area, seven players start with a ball each and three without a ball. The aim of the players without a ball is to dispossess a player with a ball. If they win the ball, they then have to avoid being dispossessed themselves. If a player dribbles out of the area, the player must switch with someone without a ball.

<u>Problem</u>: A player without a ball is more likely to be able to move quicker than a player with a ball.

<u>Solution</u>: To avoid being dispossessed, manipulate the ball using both feet and use changes of direction and speed as a response to the movement of other players.

<u>Increase levels of complexity:</u>

- Alter the ratio of players with and without a ball from 7:3 to 6:4 or 5:5.
- Decrease the size of the area.
- Add a constraint that, after 10 seconds, those in possession must swap balls with someone else who has one.
- Change the balls to a smaller sized football or a tennis ball.

Decrease levels of complexity:

- Alter the ratio of players with and without a ball from 7:3 to 8:2.
- Increase the size of the area.
- Start with one player without a ball in the square and two outside the square. After every 10 seconds, another player without a ball enters the square.

Constraint: Task is manipulated as some players act as defenders trying to dispossess an attacker with the ball affording many 1v1 situations realistic to the game.

Coaching points:

- Manipulating the ball with both feet to change direction.
- Using changes of direction and speed to evade a defender while keeping control of the ball.
- Exploiting space.
- Recognizing when to dribble and when to stay still.

Dribbling gates

Session created with SportsSessionPlanner.com

Representative design   6/10   Intentionality   8/10
Affordances   7/10   Repetition without repetition   8/10

Session organization: In a square, one player has a ball and one tries to tackle them. On each edge of the square are gates marked out by cones. Two players stay outside the square guarding the gates from the outside. The aim is to dribble through any one of the four gates without being tackled. The two outside players can guard the gates from the outside but cannot enter the square. The ball possessor can only go through an unguarded gate. Once one player dribbles through a gate without being tackled, the two on the outside go in the middle and they have a 1v1 in the middle and the practice restarts.

Problem: The defender is trying to tackle the player in possession, while the two players outside the square remove possible solutions to the attacker by guarding gates.

Solution: The player in possession must keep the ball away from the defender in the 1v1 while looking for an unguarded gate to dribble through.

Increase levels of complexity:

- Double the area size, and add another 1v1 to create more chaos adding another problem. The gates should also be made bigger if the area size is doubled.
- Once a player dribbles through a gate, the player then must try to re-enter the square through a different gate, while the three defending players try to tackle them. If a defender tackles them, the first person back in the square contests the 1v1 with them.

Decrease levels of complexity:

- Reduce the number of players guarding gates on the outside.
- Make the square bigger and the gates bigger to make it harder for the guards to cover the gates and provide more space to dribble.

Constraint: Task is manipulated by having players guarding gates on the outside and the defender in the middle combine to create many different problems requiring different solutions.

Coaching points:

- Shielding the ball by keeping their body in between the ball and the defender.
- Creating and exploiting space.
- Using quick changes of direction to attack unguarded gates.
- Shifting the ball from one foot to the other to change direction and dribble past the defender.

4v2 Rondo

Session created with SportsSessionPlanner.com

Representative design   3/10   Intentionality   7/10
Affordances   7/10   Repetition without repetition   4/10

Session organization: Four players stand in a square or diamond and try to keep the ball away from two defenders. The aim for the four players is to keep possession. If a defender wins the ball via an interception, the defender replaces the player who passed the ball. The intention is to find the free player.

Problem: The two defenders can cut off passing options by standing between the player in possession and a possible receiver.

Solution: Pass the ball to a player who is free. If positioned correctly, the two defenders will only be able to prevent passes going to two attackers leaving one free to receive a pass.

Increase levels of complexity:

- If a pass goes in between the two defenders (split pass), then they must win the ball back twice before replacing an attacker.
- Allow players to move freely in the penalty box with the aim of four players trying to retain possession from two defenders, if a defender wins it, they swap with the attacker that lost possession.

<u>Decrease levels of complexity:</u>

- Increase the area size.
- Increase the number of attackers.
- Adjust the ratio of attackers: defenders from 4:2 to 6:3 or 3:1.
- Allow players to pick up the ball and throw it to each other as the focus is to find the free player.

<u>Constraint</u>: Task is constrained by having an overload for the attackers affording them a free player to pass to.

<u>Coaching points:</u>

- Recognizing which player is free to receive a pass by looking at the movements of the defenders and the positioning of the other attackers.
- Using a soft first touch to keep the ball close to be able to pass the ball quickly when a free player becomes available to receive a pass.

Shielding the ball

Session created with SportsSessionPlanner.com

Representative design   5/10   Intentionality   7/10
Affordances   8/10   Repetition without repetition   7/10

<u>Session organization</u>: In a small space, three players start with a ball and five without a ball. The three players act as attackers and the five as defenders. The aim for players with a ball is to keep their ball for as long as possible, while those without a ball try to tackle them or force them out of the area. If a player is tackled, whoever has the ball then tries to keep it away from the defenders without a ball.

<u>Problem</u>: Players with a ball are underloaded.

<u>Solution</u>: Shield the ball from defenders by putting their body in between the ball and the defenders.

<u>Increase levels of complexity:</u>
• Decrease the area size.
• Time 15 seconds and whoever has a ball when the timer finishes gets a point.
• Repeat for a few minutes, and the winner is the player with the most points.
• Start with three players with a ball and three without, after every 20 seconds another defender joins in.
• Allow players to control more than one ball at a time.

Decrease levels of complexity:

- Increase the area size.
- Add another ball making it four players with a ball and four without.
- Award a point for retaining possession for 10 seconds, whoever has the most points after a few minutes wins.

Constraint: Task is manipulated by having five players trying to win possession off three attackers, thus affording those in possession chances to shield the ball.

Coaching points:

- Getting their body in between the ball and the defender when shielding the ball.
- Using changes of direction and speed to evade a defender while keeping control of the ball.
- Creating and Exploiting space.
- Decision-making of when to dribble into space and when to shield the ball.

In possession

Passing and scanning 2v2 + 2: General

Session created with SportsSessionPlanner.com

Representative design   5/10   Intentionality   6/10
Affordances   6/10   Repetition without repetition   7/10

<u>Session organization</u>: In a square, four players start, two from each team. In addition, two players from each team start outside the square, one on each edge. The aim of the game is to keep possession using the two outside players opposite each other. The outside players must stay on their edge of the square, and players inside the square must stay inside the square.

Introduction to the Constraints-Led Approach. https://doi.org/10.1016/B978-0-323-85026-1.00002-9

Problem: In the square, players are marked.

Solution: Use the two players outside the square to keep possession creating a 4v2.

Increase levels of complexity:

- Cannot go back to the same outside player until the other one has received a pass.
- Each pass to an outside player is a point.
- Every time a team completes 10 passes, the square gets smaller.
- Add more players into the square.
- Replace the outside players with small goals, and have a 4v4 with each team able to score in two goals.

Decrease levels of complexity:

- Add a neutral player into the middle creating an overload of 3v2 in the middle.
- Make the outside players neutral, so it becomes a 6v2 in possession.
- Make the square bigger.

Constraint: Task is manipulated to afford moments where a player scans to find space and then receives a pass. The player should then have passing options available to them.

Coaching points:

- Scanning to find space.
- Creating and exploiting space.
- Body position when receiving a pass to play quickly.
- Interactions with teammates.
- Decision-making of when to pass, dribble, or hold onto the ball.

Passing and scanning: Specific

Session created with SportsSessionPlanner.com

Representative design 7/10   Intentionality 8/10
Affordances 6/10   Repetition without repetition 8/10

Session organization: In a large area, set up a 5v5 with two neutral players located on either side of the area acting as wingers, with one player from each team acting as a striker in front of the area with a goalkeeper. The aim is to keep possession in the area using the neutral players creating a 7v5 and play the ball from the area into a striker who has 3 seconds to shoot at the goalkeeper. Play starts with a goalkeeper, and the striker is passive to begin with. Players in the middle can organize themselves into a formation.

Problem: Ball must go from the area into the striker, but all players in the area are marked.

Solution: Use the neutral players to create an overload to create space to pass forward.

Increase levels of complexity:

• Ball must go from the area to the striker within 8 seconds.
• If the striker scores, the goal is worth as many passes that took place consecutively before they scored.

- If the defending team win the ball within 6 seconds of either losing possession or the practice starting, they get a goal.
- If they score because of winning the ball within 6 seconds, that goal is worth three.

Decrease levels of complexity:

- Add in neutral players in the area creating a bigger overload in favor of the team in possession.

Constraint: Task is manipulated creating a 5v5, but the neutral-wide players offer an overload. By rewarding the defending team for winning the ball back, it will make the team out of possession work harder to win the ball quickly creating affordances for the team in possession to play forward.

Coaching points:

- Scanning regularly to find space.
- Deciding when to pass to the striker and when to keep possession.
- Decision-making of when to pass, dribble, or run with the ball.
- Creating and exploiting space.
- Interactions with teammates.
- Body position when receiving the ball.

Passing and turning: General

Session created with SportsSessionPlanner.com

Representative design 6/10   Intentionality   8/10
Affordances   9/10   Repetition without repetition   7/10

<u>Session organization:</u> In a square, set up two teams of four players and two neutral players, with four goals on each edge of the square. Teams get allocated two goals opposite each other. The aim is to score by shooting the ball into one of the goals. If the ball goes out of the area, possession turns over and resumes with a kick in. No goalkeepers allowed.

<u>Problem:</u> Two possible goals to score in.

<u>Solution:</u> Use the neutrals to create space and change the direction of the attack by turning.

<u>Increase levels of complexity:</u>

• Remove the neutrals to create a 5v5 instead of a 6v4.
• Decrease the area size.
• Players must take one or four touches; if they take two, three, or more than four, the ball turns over.
• After a team scores, one of their players joins the other team.

<u>Decrease levels of complexity:</u>

- Increase the area size.
- Make the goals bigger.
- Award one goal for every successful turn.
- Award three goals if a successful turn leads to a goal.

<u>Constraint:</u> Task is constrained with two goals to afford turning as there are two possible directions to attack in. The additional rule of one or four touches will afford chances to turn by letting the ball run across their body without touching it.

<u>Coaching points:</u>

- Turning to utilize the two goals and change direction.
- Turning by letting the ball run across your body without touching it.
- Creating and exploiting space.
- Interactions with teammates.
- Decision-making of when to pass, dribble, turn, or shoot.
- Body position when receiving the ball.

Creating 3v2 situations: General

Session created with SportsSessionPlanner.com

Representative design   6/10   Intentionality   7/10
Affordances   7/10   Repetition without repetition   8/10

Session organization: Divide a pitch into thirds. Set up a 3v2 in the starting third, leave the middle third empty, and have a 2v2 with a goalkeeper in the final third. The aim is to start in the starting third and try to score a goal in the final third. One player from the starting third can run through the middle third when the ball is played forward. One player from the final third can drop into the middle third to receive unopposed, but they can only be in the middle third for 3 seconds at a time. Defenders are not allowed out of their third. If the defenders win the ball, they aim to score in the mini goal and they must stay in their thirds.

Problem: Players are not given a 3v2 situation to begin with, they have to problem-solve to create it.

Solution: Once a pass is played forward, an attacker can make a run into the final third to create a 3v2. In addition, an attacker has to time when to drop short to receive a pass in the middle third.

<u>Increase levels of complexity:</u>

- Attackers have 5 seconds in the final third to get a shot away.
- Allow a supporting defender to enter the final third creating a 3v3; after 5 seconds, another attacker can arrive creating a 4v3 and continue until all the players are in the final third.

<u>Decrease levels of complexity:</u>

- Allow attackers to dribble through the middle third to create a 3v2 in the final third.
- Divide the pitch in half removing the middle third with the new rule of a maximum of three attackers and two defenders in any half at any time.

<u>Constraint:</u> Task is manipulated to afford chances to create 3v2 situations.

<u>Coaching points:</u>

- Timing of the movement of the ball either running into the final third or dropping short to receive.
- Timing and weight of pass in a 3v2 situation.
- Decision-making of when to shoot, pass, or dribble.
- Creating and exploiting space.
- Interactions with teammates.

Playing out with a back four: Specific

Session created with SportsSessionPlanner.com

Representative design   8/10   Intentionality   8/10
Affordances   6/10   Repetition without repetition   8/10

Session organization: In one half, the aim is for a back four and two midfielders to play the ball out from the goalkeeper and dribble through a gate on the halfway line. The opposition have four players to defend and apply pressure, and if they win the ball back, they can attack.

Problem: Gates are on the halfway line.

Solution: Create space by spreading out making it difficult for the four defending players to win the ball.

Increase levels of complexity:

- Add in two more defenders to create a 7v6.
- After every 5 seconds add another defender.
- Defending team get one goal every time they win the ball and shoot.
- Attacking team get two goals if they play through a gate within 8 seconds.
- At least one pass from the attacking team must be in the air.

<u>Decrease levels of complexity:</u>

- Start with two defenders and add one more every 5 seconds.
- Bring the gates closer.
- Defending players are not allowed in the penalty area.

<u>Constraint:</u> Task is manipulated to afford passing out from the back by only having four defenders against seven attackers Furthermore, having to dribble through a set of cones on the halfway line gives the attackers direction and also forces them to keep control of the ball. Time constraints being added will afford quicker passing and a higher intensity.

<u>Coaching points:</u>

- Creating and exploiting space.
- Body position when receiving a pass.
- Using the goalkeeper to add depth.
- Varying the distance of passes.
- Decision-making of when to pass, dribble, shoot, or run with the ball.
- Interactions with teammates.

Playing out on one side: Specific

Session created with SportsSessionPlanner.com

Representative design   6/10   Intentionality   8/10
Affordances   8/10   Repetition without repetition   7/10

Session organization: On one side of half a pitch, set up one goalkeeper, one full back, two center backs and two midfielders, and four opposition players defending. The aim of the attackers is to dribble through one of the gates near the halfway line, one near the touchline and one closer to the center. If the defenders win the ball, they can attack.

Problem: Players start on the side of the pitch where the gates are located.

Solution: Create space using the overload to play the ball up to the halfway line.

Increase levels of complexity:

- Add in a defender so it is a 6v5.
- Attackers have to reach a gate within 10 seconds.
- Every pass must be in the air.
- After 5 seconds an extra defender is added continuously.
- If the defending team win the ball, they have 6 seconds to shoot.
- Defending team get a goal if the ball goes out of play off the attacking team.

Decrease levels of complexity:

- Remove a defender.
- Start with two defenders and add one every 5 seconds.
- Bring the gates closer.
- Defending players are not allowed in the penalty area.

Constraint: Task is manipulated by starting players from one side of the pitch with the gates ahead of them to afford building up the play on one side. Ariel passes can be introduced to afford quicker passes over a longer distance. Time limits will also increase the intensity and speed of the play.

Coaching points:

- Creating and exploiting space.
- Body positioning when receiving a pass.
- Decision-making of when to pass, dribble, or run with the ball.
- Interactions with teammates.
- Varying the types of pass.

Playing through central midfield: Specific

Session created with SportsSessionPlanner.com

Representative design   8/10   Intentionality   9/10
Affordances   8/10   Repetition without repetition   8/10

Session organization: Mark out an area in between the halfway line and the penalty area. Set up a 2v2 acting as central midfielders in the area with three players acting as feeders based on the halfway line with a striker, defender, and goalkeeper in the penalty area. The aim of the attacking team is to play from a feeder through the area to the striker who has 5 seconds to shoot. If the defending team win the ball, they get a goal for every successful pass to a feeder.

Problem: Midfielders are marked, and they have to receive and try to pass to the striker.

Solution: Attackers must find space to receive a pass and then try to pass forward to the striker.

Increase levels of complexity:

- Attackers have 6 seconds for the ball to go from the feeders to the striker having a shot.
- Decrease the area size in the middle.
- Add defenders to pressure the feeders.
- If the two defenders win the ball within 5 seconds and pass to a feeder, they get three goals.

- After a goal scored by the attackers, the area gets smaller.
- After a goal scored by the defenders, the area gets bigger.

<u>Decrease levels of complexity:</u>

- Remove a defender from the middle creating a 2v1 in favor of the attackers.
- The attackers get a goal for every successful pass to the striker.

<u>Constraint:</u> Task is manipulated to afford the attacking team to play through central midfielders. Time limits will afford for intensity and increase the speed of play.

<u>Coaching points:</u>

- Interactions between teammates.
- Creating and exploiting space.
- Body positioning when receiving a pass.
- Decision-making of when to pass, dribble, shoot, or run with the ball.
- Encouraging creativity and innovation.

Switching play 10v8: General

Session created with SportsSessionPlanner.com

Representative design   8/10   Intentionality   8/10
Affordances   7/10   Repetition without repetition   7/10

Session organization: In an area larger than half a pitch, set up a 10v8. The aim of the 10 attackers is to score in a mini goal in the wide areas while eight defenders try to win possession and attack quickly into a full-sized goal centrally.

Problem: Mini goals are located in the wide areas offering direction.

Solution: Use the overload to force the defending team to herd toward the ball on one side before switching the play to the opposite side where free players should be located.

Increase levels of complexity:

- Defending team have 6 seconds to shoot when they gain possession; once the 6 seconds are up, the attacking team regain possession wherever the ball is.
- A goal scored following a switch of play is worth three.
- If the defending team win the ball in the wide areas, they get a goal; if they score from winning the ball in the wide area, that is worth three.

Decrease levels of complexity:

- Any time the attacking team have passing options on both sides of the pitch, they get one goal.
- A switch of play from the attacking team is worth two goals, if the switch leads to a goal it is worth 3.

Constraint: Task is manipulated with the overload and a wide pitch for attackers to maintain width affording them chances to switch the play. Task is further manipulated by the location of the goals in wide areas. Time constraints will also lead to the defending team pressing in wide areas and attacking quickly.

Coaching points:

- When and how to switch the play.
- Maintaining width and depth.
- Creating and exploiting space.
- Decision-making of when to pass, shoot, dribble, or run with the ball.
- Interactions between teammates.

Switching play 9v9: SSG

Session created with SportsSessionPlanner.com

Representative design   9/10   Intentionality   8/10
Affordances   6/10   Repetition without repetition   7/10

Session organization: In an area larger than half a pitch, set up a 9v9 including goalkeepers. The aim is to score in the full-sized goals located centrally. The team in possession must keep two players in each wide area at all times when in possession. Wide areas can be marked out using the width of the penalty box if required.

Problem: Attacking team need to keep two players on each side in wide positions when in possession.

Solution: Try to draw the defending team onto one side of the pitch before switching it to the other side where two players are located.

Increase levels of complexity:

- A goal scored from a switch of play is worth three.
- If a team score within six passes of winning the ball, the goal is worth three; after six passes, normal play resumes.

- If a team switches the play and scores within 6 passes, that goal is worth six.
- Defending team get a goal if they win possession in the wide areas.

Decrease levels of complexity:

- Every switch of play is worth one goal.
- Make two players neutral creating a 10v8 and making it easier to keep two players in each side of the pitch in wide areas when in possession.

Constraint: Task is manipulated as the pitch is shorter but still offers full width. Task is further constrained by forcing them to keep four players in wide areas in possession and rewarding switches of play. Combining these will afford players many chances to switch the play.

Coaching points:

- When and how to switch the play.
- Maintaining width and depth.
- Creating and exploiting space.
- Decision-making of when to pass, shoot, dribble, or run with the ball.
- Interactions between teammates.
- Body position when receiving the ball.

Diagonal passing 8v8: SSG

Session created with SportsSessionPlanner.com

Representative design   8/10   Intentionality   7/10
Affordances   7/10   Repetition without repetition   7/10

<u>Session organization:</u> In a large area, set up an 8v8 with goalkeepers, with the corners of the pitch removed. The aim is to score in a full-sized goal, and normal rules apply with throw-ins being taken on the new touchlines even in the corners.

<u>Problem:</u> The corners of the pitch are removed so there is less width and depth to use when in possession.

<u>Solution:</u> Create angles for diagonal passes to be played using the width when possible. The most width can be found in the middle of the pitch.

<u>Increase levels of complexity:</u>

- A goal scored within five passes is worth three.
- Make the pitch longer to afford more vertical passes maintaining the removal of the corners.
- If a team scores within 6 seconds of winning the ball back, that goal is worth three.
- After any team scores, one of their players joins the other team.

Decrease levels of complexity:

- A goal is awarded for every three diagonal passes made without losing the ball.
- Make two players neutral creating an overload 10v8 to afford more passing options.
- A goal scored as a result of a diagonal pass is worth three.

Constraint: Task is manipulated by the pitch dimensions to afford fewer options to pass sideways and more options diagonally as there is less width and depth.

Coaching points:

- Weight and type of pass.
- Interactions between teammates.
- Creating and exploiting space.
- Decision-making of when to pass, dribble, shoot, or run with the ball.
- Body position when receiving the ball.
- Encouraging creativity and innovation.

Attacking with an underload: Specific

Session created with SportsSessionPlanner.com

Representative design 8/10   Intentionality   9/10
Affordances   7/10   Repetition without repetition   6/10

<u>Session organization:</u> In half a pitch, set up a 6v9 underloading the attacking team with mini goals on the halfway line. The aim of the attacking team is to try to score in the full-sized goal centrally, and the defending team try to play into mini goals on transition.

<u>Problem:</u> Score despite having less players than the opposition and being aware of preventing counterattacks on transition.

<u>Solution:</u> Move off the ball to find space and use quick and incisive passing to disorganize the defense. Dribbling will also draw defenders toward one attacker to create 1v1 situations in other areas of the pitch.

<u>Increase levels of complexity:</u>

- Make the pitch smaller.
- Reduce the ratio of attackers to defenders.
- Add a time limit for the attackers to shoot.

Decrease levels of complexity:

- Reduce the number of players so there is more space for attackers.
- After a goal scored by the defenders, one player joins the attacking team. If the attackers score next, one player goes back on the defending team.

Constraint: Manipulate the task by adjusting the ratio of defenders to attackers to create an underload. The task is further manipulated by adding two goals for the defenders to give them a target on transition.

Coaching point:

- Move off the ball to find space.
- Quick and incisive passing.
- Interchanging positions to disorganize the defense.
- Interactions with teammates.
- Body position when receiving the ball.
- Encouraging creativity and innovation.

Overloads in wide areas: Specific

Session created with SportsSessionPlanner.com

Representative design   9/10   Intentionality   9/10
Affordances   6/10   Repetition without repetition   6/10

Session organization: In half a pitch, set up a 7v7. The aim is to use a 2v1 over-load in wide areas to get crosses into the box for the other attackers. Every goal scored from a cross is worth three. One of the two central midfielders starts the prac-tice. If a defender wins possession, the defender's team aim to progress the ball un-der control to the halfway line for one goal.

Problem: The only place where the attacking team have an overload is in the wide areas.

Solution: Play the ball wide creating a 2v1 situation to get crosses into the pen-alty area for the striker and other attackers.

Increase levels of complexity:

- Attacking team have 6 seconds from the start of play to deliver a cross.
- Add in extra defenders.
- If the defenders win the ball within 6 seconds, that is worth one goal.
- If the defending team win possession and get the ball to halfway under control, they get three goals.

Decrease levels of complexity:

- Remove a defender a creating a 7v6 affording more 2v1s or 3v2s to get crosses into the box.
- Every cross is worth a goal.

Constraint: Task slightly manipulated by the setup and the reward for scoring from a cross to afford overloads in wide areas to provide opportunities for crosses.

Coaching points:

- Different types of crosses based on the movement of teammates in the penalty area.
- Interactions between teammates.
- Creating and exploiting space.
- Decision-making of when to pass, dribble, cross, shoot, or run with the ball.
- Body position when receiving the ball.
- Encouraging creativity and innovation.

Using wide areas: SSG

Session created with SportsSessionPlanner.com

Representative design   7/10   Intentionality   8/10
Affordances   6/10   Repetition without repetition   8/10

Session organization: In half a pitch, set up an 8v8+2 neutrals. The aim is to score in a full-sized goal on halfway. Neutrals must stay within the width of the box. When in possession, teams must have at least one player in each wide area between the touchline and the width of the penalty box. Teams can have more than one player in wide areas.

Problem: Must have two players in wide areas when in possession.

Solution: Use the wide players to spread out the defense to create space to attack.

Increase levels of complexity:

- A goal scored by using the wide players is worth three. It is possible to use the wide player without giving them the ball. It is up to the coach to decide.
- If the defending team score within 6 seconds of winning the ball back, that goal is worth three.
- Players can either take one or four touches. If a player uses two, three, or more than four, the ball turns over to the opposition where it is.

Decrease levels of complexity:

- Neutrals stand in the wide areas all the time.
- Every cross is worth one goal.
- A goal scored from a cross is worth three.

Constraint: Task is constrained by adjusting the numbers with neutrals creating a 10v8 in possession and the forcing of two players to maintain width to afford wide play and/or central overloads. Adding the one or four touch constraints will also afford more combination play.

Coaching points:

- Creating and exploiting space.
- Interactions between teammates.
- Types of crosses.
- Decision when to pass, cross, dribble, shoot, or run with the ball.
- Combination play.
- Body position when receiving a pass.

Attacking final third: Specific

Session created with SportsSessionPlanner.com

Representative design   6/10   Intentionality   8/10
Affordances   8/10   Repetition without repetition   8/10

Session organization: In half a pitch, set up a 6v6. The aim of the attacking team is to try to score. If the defending team win possession, they try to score in a mini goal on halfway. Play starts on the halfway line with one central midfielder. The players in the attacking team can either take one or four touches outside the penalty area. If an attacker uses two, three, or more than four, the ball turns over to the where it is.

Problem: Players can only take one or four touches unless they are in the penalty area.

Solution: Use the overload to get into the penalty area where unlimited touches are allowed.

Increase levels of complexity:

- Add an extra attacker.
- If the defending team win the ball within 8 seconds and score, that goal is worth three.
- If the attacking team score by using passing combinations, that goal is worth three.
- After a team scores, one of their players joins the other team.

Decrease levels of complexity:

- Remove a midfielder from each team so the only remaining midfielder becomes like an unofficial feeder for the front four.
- Attackers get a goal for entering the penalty area.

Constraint: Task is manipulated using the one or four touch constraints to afford quick passing combinations.

Coaching points:

- Decision-making of when to dribble, pass, or shoot.
- Encouraging creativity and innovation.
- Interactions between players.
- Creating and exploiting space.
- Combination play.

Creating and scoring: Specific

Session created with SportsSessionPlanner.com

Representative design   6/10   Intentionality   8/10
Affordances   8/10   Repetition without repetition   8/10

Session organization: Within the width of the box, set up a 3v3 with two feeders and a goalkeeper. The aim of the attacking team is to try to score. Feeders start with the ball and play it to the three attackers who try to create and score. One feeder can support the attack, and the feeders can be tackled. If a defender wins the ball, the defender's team play it to the feeder for one goal, and the practice restarts. Defending team also get a goal for winning the ball within 6 seconds. This can be altered easily to suit different numbers.

Problem: Defenders get a goal by winning the ball within 6 seconds and another for passing successfully to a feeder.

Solution: Use the feeders to create an overload and exploit spaces.

Increase levels of complexity:
- Add another defender every 15 seconds.
- Attacking players can either take one or four touches. If an attacker uses two, three, or more than four, the ball turns over to the opposition where it is.
- After the attacking team score a goal, add another defender.

Decrease levels of complexity:

- Remove a defender.
- Remove the reward for the defending team winning the ball within 6 seconds.

Constraint: Task is manipulated by rewarding the defenders for winning possession quickly to afford moments for attackers to create and score. The additional one or four touch constraints will afford more combination play.

Coaching points:

- Interactions between teammates.
- Body position when receiving a pass.
- Creating and exploiting space.
- Combination play.
- Decision-making of when to pass, shoot, dribble, or run with the ball.
- Encouraging creativity and innovation.

3v3 combination play: SSG

Session created with SportsSessionPlanner.com

Representative design   6/10   Intentionality   7/10
Affordances   8/10   Repetition without repetition   7/10

Session organization: In a small space, set up a 3v3 with no goalkeepers. The aim is to score in a mini goal. The only additional rule is that players must either use one or four touches when in possession. If a player uses two, three, or more than four touches, the ball turns over to the opposition where it is.

Problem: Players either have to pass/shoot on their first touch, or they must take three touches before passing or shooting.

Solution: Look around (scan) before receiving a pass to inform the decision whether to pass/shoot using their first touch or control the ball.

Increase levels of complexity:

• Increase numbers to a 4v4 or 5v5, add goalkeepers.
• Increase the area size both width and depth.

Decrease levels of complexity:

- Allow up to four touches in possession.
- If a goal is scored after two passes, it is worth two; after three passes, it is worth three and so on.

Constraint: Task is manipulated using the one or four touch constraints to afford quick passing and movement.

Coaching points:

- Scanning before receiving the ball.
- Timing movements off the ball to be able to receive a pass.
- Decision-making of when to shoot, pass, dribble, or run with the ball.
- Body position when receiving the ball.

Shooting: SSG

Session created with SportsSessionPlanner.com

Representative design   7/10   Intentionality   8/10
Affordances   9/10   Repetition without repetition   7/10

<u>Session organization:</u> On a small pitch, set up a 6v6 with three attackers in one half and two defenders in the other half for each team creating two 3v2 situations. Players must stay in their halves. The aim is to score in a full-sized goal past a goalkeeper.

<u>Problem:</u> Attackers have a 3v2 overload, and the pitch is small.

<u>Solution:</u> Use the overload to create space to shoot as much as possible.

<u>Increase levels of complexity:</u>

• If the ball has been in one half for 6 seconds, another attacker can enter creating a 4v2.
• If the defenders win the ball and their team scores within 6 s, that goal is worth three.
• After every goal, the pitch is made 4 yards wider (2 yards on each side).

<u>Decrease levels of complexity:</u>

- Increase the area width but not the length so the goals are still close together giving players more space to get shots away.
- Make two players neutral and allow players to move freely creating a 5v5+2.

<u>Constraint:</u> Task is manipulated to create overloads, and a short pitch affords players more opportunities to shoot at goal.

<u>Coaching points:</u>

- Different shooting techniques.
- Focusing on the connection with the ball.
- Decision-making of when to shoot, pass, dribble, or run with the ball.
- Creating and exploiting space.
- Interactions between teammates.

Attacking third man runs: SSG

Session created with SportsSessionPlanner.com

Representative design   6/10   Intentionality   9/10
Affordances   9/10   Repetition without repetition   9/10

Session organization: In an area, set up a 5v5 but turn the goals around to face the opposite direction. The aim of the game is to score past the goalkeeper in a full-sized goal. Attackers must make third man runs off the ball to receive a pass behind the goal line, which doubles up as an offside line. Attacker then has a 1v1 against the goalkeeper. Defenders cannot go beyond the goal line.

Problem: Cannot score without making a third man run off the ball.

Solution: Make a third man run off the ball into the space beyond the goal line.

Increase levels of complexity:

- Increase the number of players.
- Increase the pitch size.
- Players have 3 seconds from receiving the ball beyond the goal line to shoot.
- If a team scores within 6 seconds of winning the ball, that goal is worth three.

<u>Decrease levels of complexity:</u>

• Add in neutral players to offer more opportunities where someone will have time in possession.
• After a team scores, one of their players joins the other team.

<u>Constraint:</u> Task is manipulated by turning the goals around and adjusting the pitch size. Neutral players will give players more time in possession, which can afford more runs to be made.

<u>Coaching points:</u>

• Speed and direction of the run.
• Interactions with teammates.
• Decision-of when to pass and when to make a third man run.
• Creating and exploiting space.

Crossing and finishing: SSG

Session created with SportsSessionPlanner.com

Representative design   6/10   Intentionality   8/10
Affordances   9/10   Repetition without repetition   7/10

Session organization: In half a pitch mark, out a small area in the middle. Set up each team with one goalkeeper with three defenders in front of them. In addition, set up a 2v2 in the middle area marked out and two neutral players in wide positions, one player from the middle must stay in the middle, and the neutral players must stay wide. The aim of the practice is to create chances for the striker using the neutral players as wingers. Play begins from a goalkeeper who can play it to anyone.

Problem: Striker is overloaded 3v1.

Solution: Play the ball wide to a neutral player who can go forward and cross to a striker, midfielder, or the other neutral player.

Increase levels of complexity:

- Remove the middle square so both teams have a back three, two midfielders, and one striker.
- The ball must go through the middle area before a goal can be scored.
- All players except the neutrals must use either one or four touches.

Decrease levels of complexity:

- Change a defender to attacker in both teams creating two 2v2 situations instead of two 3v1 situations.
- After every goal scored, the middle area is made bigger.

Constraint: Task is manipulated to afford moments when the ball is played wide and can be crossed in.

Coaching points:

- Different types of crosses based on the movement of teammates in the penalty area.
- Interactions between teammates.
- Creating and exploiting space.
- Body position when receiving the ball.
- Decision-making of when to pass, cross, dribble, shoot, or run with the ball.

Transitions

Counter pressing: General

Session created with SportsSessionPlanner.com

Representative design   6/10   Intentionality   7/10
Affordances   7/10   Repetition without repetition   7/10

<u>Session organization</u>: Mark out one large rectangle, and in the middle of the rectangle, mark out a small square. Set up three attackers and one defender with the aim to keep possession. Three defenders start outside the square. When the defender in the square wins possession, the defender aims to play it to a teammate outside the square, and the original attackers become defenders aiming to win the ball back in the large rectangle. When the three win possession back, they aim to get the ball back into the small square; if they do one opposition, player can defend in the square, and the practice continues.

Introduction to the Constraints-Led Approach. https://doi.org/10.1016/B978-0-323-85026-1.00011-X

Problem: The three attackers are underloaded.

Solution: On defensive transition, press aggressively and collectively to regain possession.

Increase levels of complexity:

- Make the rectangle size smaller.
- Defending team have 6 seconds to win the ball after losing it.
- Every time the one defender wins the ball, the square gets bigger.

Decrease levels of complexity:

- Allow two defenders against the three in the smaller square.
- Adjust the rules so players throw the ball like netball, and only interceptions can regain possession.

Constraint: Task is manipulated to afford lots of instances of transitions due to the size of the square and positioning of players outside the square.

Coaching points:

- Pressing as a unit.
- Interactions between teammates.
- Trigger movements.
- Decision-making of when to press the ball and when to cut passing lanes.

Passing on transitions: General

Session created with SportsSessionPlanner.com

Representative design   7/10   Intentionality   9/10
Affordances   8/10   Repetition without repetition   8/10

Session organization: In a large rectangle, set up a 5v2 with three defenders stood outside the rectangle, with two mini goals located opposite each other outside the rectangle. The aim of the five attackers is to keep the ball for 20 s and then shoot at a mini goal. If a defender wins the ball, the three spare defenders join in and three players who were previously attacking drop out creating a 5v2 in favor of the other team, who aim to keep possession for 20 seconds and then shoot.

Problem: Must win the ball back underloaded within 20 seconds and then pass to a teammate outside the rectangle.

Solution: Try to win the ball back near a teammate outside the rectangle, so the first pass on transition is easier.

Increase levels of complexity:

• After every 5 seconds, a new defender is added, but after every transition, it must go back to a 5v2 to begin.
• Players must take one or four touches; if they take two, three, or more than four, the ball turns over.

Decrease levels of complexity:

- Adjust it to a 5v3 with two on the outside.
- Attacking team must use either one or four touches.

Constraint: Task is manipulated due to the 5v2 to afford moments of chaos on transition, and the one or four constraints will also afford more transitions.

Coaching points:

- Having two players close to the ball when in possession so on defensive transition they become the ones who press the ball aggressively.
- Using pressing triggers.
- Interactions between teammates.
- Decision-making of when to press the ball and when to cut passing lanes.

3v3 defensive transition: SSG

Session created with SportsSessionPlanner.com

Representative design   6/10   Intentionality   7/10
Affordances   8/10   Repetition without repetition   7/10

<u>Session organization</u>: In a small area, set up a 3v3 with no goalkeepers, and the aim is to score in a mini goal. The only additional rule is that if a team scores a goal for winning the ball back within 6 seconds, that is worth three.

<u>Problem</u>: Goals are worth more if they come from winning the ball back within 6 seconds of losing it.

<u>Solution</u>: Counterpress aggressively as a reaction to losing the ball.

<u>Increase levels of complexity</u>:

• Increase the area size both depth and width after every goal.
• Increase numbers to a 4v4.

<u>Decrease levels of complexity</u>:

• Reduce to a 2v2.
• Add a rule of 3 seconds on the ball per player to increase the amount of times a player loses possession.

Constraint: Task is manipulated with the rule that if you get a goal for winning the ball back within 6 s of losing it and if you score from that it is worth 3.

Coaching points:

- React quickly to losing possession.
- Herd toward the ball for 6 seconds after losing it but go into a more conventional shape to defend after the 6 seconds.
- Force the opposition wide and backward in 1v1 situations.
- Decision-making of when to press the ball and when to cut passing lanes.
- Interactions between teammates.

3v3 attacking transition: SSG

Session created with SportsSessionPlanner.com

Representative design 6/10   Intentionality 7/10
Affordances 8/10   Repetition without repetition   7/10

Session organization: In a small area, set up a 3v3 with no goalkeepers, and the aim is to score in a mini goal. The only additional rule is that players have to shoot within 6 seconds of winning possession, and if they score within 6 seconds, that goal is worth three. If a goal is scored not following a transition (from a restart), that goal is worth one.

Problem: Players have to shoot within 6 seconds of winning the ball.

Solution: Attack quickly and look to win possession as close to oppositions goal as possible.

Increase levels of complexity:
- Increase numbers to a 4v4.
- Increase the area size both depth and width after every goal.

<u>Decrease levels of complexity:</u>

- Change to a 2v2.
- Add a rule that players must keep one player in each half at all times.

<u>Constraint</u>: Task is manipulated by the rule that you have to shoot within 6 seconds of winning possession, and if you score within 6 seconds, it is worth three.

<u>Coaching points:</u>

- Decision-making of when to dribble, pass, shoot, or run with the ball.
- Attacking at speed.
- Interactions between teammates.
- Creating and exploiting space.

Counter attacking 4v4+2: SSG

Session created with SportsSessionPlanner.com

Representative design   7/10   Intentionality   7/10
Affordances   8/10   Repetition without repetition   7/10

Session organization: In a large area, set up 5v5+2 neutrals, who play for the team in possession with goalkeepers and full-sized goals. The only additional rules are that when a team wins the ball, every pass must go forward. And if a team score, they keep possession of the ball, and play restarts from their goalkeeper.

Problem: Every pass must go forward on attacking transition.

Solution: Run beyond the player in possession on transition to be a passing option and stretch the opposition.

Increase levels of complexity:

- Attackers have 6 seconds to score; otherwise, the ball turns over.
- For a goal to count, all outfield players must be in the opposition half.

Decrease levels of complexity:

- A goal scored within 6 seconds from a team winning the ball is worth three.
- One neutral must stay in each half.

Constraint: Task is manipulated with additional rules on the counterattack and having neutral players creating overloads.

Coaching points:

- Neutral players stretching the opposition.
- Decision-making of when to pass, shoot, dribble, or run with the ball.
- Interactions between teammates.
- Creating and exploiting space.
- Attacking at speed.

Counter attacking 6v6: SSG

Session created with SportsSessionPlanner.com

Representative design 6/10   Intentionality   9/10
Affordances   9/10   Repetition without repetition   7/10

<u>Session organization</u>: On a long but narrow pitch, set up a 6v6 with goalkeepers and full-sized goals. The only additional rule is that a goal only counts if all team members (except the goalkeeper) are in the opposition half.

<u>Problem</u>: Everyone must be in the opposition half for a goal to count.

<u>Solution</u>: On transition, counterattack as a team into space.

<u>Increase levels of complexity</u>:

- On attacking transition, every pass must go forward.
- Attackers must either take one or four touches.
- After a team scores a goal, one of their players joins the other team.
- Counterattacking goals are worth three.

<u>Decrease levels of complexity</u>:

- Decrease the area size.
- Add a neutral player.
- If a team scores a goal within 6 seconds of winning the ball, that goal is worth three.

Constraint: Task is manipulated as goals only count if everyone is in the opposition half, thus affording moments to counterattack as there will be space left by the opposition.

Coaching points:

- Creating and exploiting space.
- Interactions between teammates.
- Decision-making when to pass, shoot, dribble, or run with the ball.
- Secure the first pass on the transition.
- Runners going ahead of the ball.

Out of possession

9

Defending central areas: Specific

Session created with SportsSessionPlanner.com

Representative design   7/10   Intentionality   8/10
Affordances   7/10   Repetition without repetition   7/10

<u>Session organization</u>: In an area, the width of the penalty box, set up a 4v4 and one goalkeeper with one feeder who starts the play. The aim of the feeder is to play the ball to an attacker, and the defenders aim to win the ball. The attackers aim to score in the full-sized goal while the defenders try to win possession and play the ball to the feeder for a goal. Attackers can only enter the penalty area by dribbling past a defender. If the ball goes out of the area, the defenders get one goal.

<u>Problem</u>: Attackers can only enter the penalty area by dribbling.

Introduction to the Constraints-Led Approach. https://doi.org/10.1016/B978-0-323-85026-1.00004-2

91

Solution: Defend the edge of the penalty area.

Increase levels of complexity:

- If the defenders win the ball and play it to the feeder within 3 seconds, they score three.
- Defenders must win the ball back within 6 seconds.

Decrease levels of complexity:

- Remove an attacker creating a 4v3 overload for defenders.
- Every time the attackers scores they lose a player.

Constraint: Task is manipulated to afford defenders moments to defend 1v1 as part of a defensive unit.

Coaching points:

- Interactions between teammates.
- Trying to keep chest on to attackers in 1v1 situations.
- Decision-making of when to jockey, tackle, or intercept.
- Trying to stay chest on in 1v1 situations with an attacker.

Defending the final third: Specific

Session created with SportsSessionPlanner.com

Representative design   8/10   Intentionality   8/10
Affordances   6/10   Repetition without repetition   7/10

Session organization: In half a pitch, set up a 5v6 including a goalkeeper. Cones mark out three gates on the halfway line. The aim for the defenders is to win the ball and pass the ball through a gate on halfway. The attackers aim to score in a full-sized goal protected by the goalkeeper. Play starts in front of one gate with an attacker passing the ball. If the ball goes out of play, play restarts from the halfway line again with the attackers.

Problem: The attackers start with the ball, and the gates are on the halfway line.

Solution: Win the ball back as close to the halfway line as possible.

Increase levels of complexity:

• The defending team have 8 seconds to win the ball.
• After 5 seconds, a new attacker is added continuously.
• Once the defenders win the ball, every pass must go forward.

Decrease levels of complexity:

• Remove an attacker, leading to a 6v4 in favor of defenders.
• Every time the attackers score, one player joins the defending team.

Constraint: Task is manipulated by the number of players to afford chances for the defense to work together as a unit.

Coaching points:

- Interactions between teammates.
- Covering defenders if they step out to press the ball.
- Trying to keep chest onto attackers in 1v1 situations.
- Decision-making of when to jockey, tackle, or intercept.
- Force the attacker wide and backward in 1v1 situations.
- Large strides to cover big spaces but short strides when dealing with a 1v1 situation.

Defensive covering: Specific

Session created with SportsSessionPlanner.com

Representative design   6/10   Intentionality   7/10
Affordances   9/10   Repetition without repetition   7/10

<u>Session organization</u>: In half a pitch, set up a back four and one goalkeeper against three attackers and one feeder. A defender is told to press the attacker in possession and leave space behind them. The aim of the feeder is to play the ball into an attacker, and a defender pushes up close to him as there are no midfielders in front. The attacking team aim to score in the full-sized goal protected by the goalkeeper. When defenders win the ball, they play it to the feeder for a goal. It is important that the defenders do not have any defensive midfielders in front of them.

<u>Problem</u>: Defenders have no protection in front of them.

<u>Solution</u>: When a defender pushes out the other, three must move to cover the space left behind.

<u>Increase levels of complexity</u>:
- Add an extra attacker after 4 seconds.
- If the defending team score, a defender joins the attacking team.

<u>Decrease levels of complexity</u>:
- Remove an attacker to create a 4v2.
- Narrow the area to only use the width of the box.

<u>Constraint</u>: Task is constrained to afford moments where defenders must cover behind a defender who is putting pressure on the ball.

<u>Coaching points:</u>

- Interactions between teammates.
- Decision-making of when to press the ball and when to cover space left behind.
- Trying to stay chest onto the attacker in 1v1 situations.
- Force the attacker wide and backward in 1v1 situations.

Defending 3v2 situations: General

Session created with SportsSessionPlanner.com.

Representative design 6/10 Intentionality 8/10
Affordances 9/10 Repetition without repetition 8/10

Session organization: In one half, set up a 3v2 with additional two defenders and three attackers on halfway. The aim of the defending team is to win possession and pass to a player on halfway. The attacking trios take it in turn to attack the two defenders and try to shoot at the goalkeeper. If the two defenders can delay the opposition for 3 s, two more defenders join to support.

Problem: Defenders are overloaded in a 3v2 situation in a large space.

Solution: Delay the attackers so the two supporting defenders can join creating a 4v3 situation.

Increase levels of complexity:

• After 6 seconds, another three attackers get added creating a 6v4.
• Attackers only have 6 seconds to shoot in a 6v4 situation.

Decrease levels of complexity:

- Position an extra defender 10 yards in front of the two, and they are active as soon as the ball goes past them.
- Begin with two attackers, and after 3 seconds, an extra attacker joins.

Constraint: Task is manipulated as defenders know if they can delay the attackers, they get rewarded with two more defenders. But there could be more attackers joining if they do not win the ball quickly.

Coaching points:

- Large strides to cover big spaces but short strides when dealing with a 1v1 situation.
- Trying to stay chest on with the attacker in a 1v1 situation.
- Decision-making of when to delay and when to tackle or intercept.
- Interactions between teammates.
- Force the attacker wide and backward in 1v1 situations.

Defending as a midfield unit: General

Session created with SportsSessionPlanner.com

Representative design   8/10   Intentionality   8/10
Affordances   7/10   Repetition without repetition   8/10

Session organization: Mark out three areas. The aim of the defenders in the middle area is to intercept passes being played from the attackers in front of them to the attackers behind them. If a pass is made through the line of defenders, the game continues; if a defender intercepts or a pass goes out of the area, the attackers who played the pass switch with the defenders. To begin, players must stay in their area. Ball cannot go over head height.

Problem: Potential receivers behind the line of defenders.

Solution: Shuffle as a unit to reduce gaps in between players to intercept.

Increase levels of complexity:
- One of the defenders can press the ball and exit the area at a time.
- More than one defender can leave their area to either press the ball or block a pass to an intended receiver.
- A pass through is one goal for the group of attackers who played the pass, the group with the most goals after a few minutes wins.
- Ball can only be in one area for a maximum of 5 seconds.

Decrease levels of complexity:

- Decrease the width of the areas so there is less space for defenders to cover.
- Divide the areas in half creating two identical practices side by side with only two players in each area instead of four.

Constraint: Task has been manipulated by using the areas to constrain where players can move to afford interceptions.

Coaching points:

- Interactions between teammates.
- Defenders checking their shoulders.
- When one defender presses the man in possession, the other covers the space left behind.
- Defenders shuffling sideway staying compact.

Defending in midfield: SSG

Session created with SportsSessionPlanner.com

Representative design   6/10   Intentionality   8/10
Affordances   8/10   Repetition without repetition   6/10

<u>Session organization</u>: On half a pitch within the width of the box, divide the space into thirds. Set up a 6v6, both teams use a 1-3-1 formation staying in their third. If a midfielder plays the ball to the striker, all the midfielders can break into the attacking third to support creating an overload.

<u>Problem</u>: Defender overloaded when midfielders enter the final third.

<u>Solution</u>: Work together as a midfield three to cut out passes to the striker.

<u>Increase levels of complexity</u>:

- As well as midfielders advancing into the attacking third, a midfielder can help their defender after 3 seconds.
- When the midfielders advance, the defender can push up into the middle third.
- After every goal scored, make the pitch narrower.

<u>Decrease levels of complexity</u>:

- Only the midfielder who passes to the striker can enter the final third.
- Attackers have a maximum of five passes to find the striker.
- Attackers only have 6 seconds in the final third to shoot.

<u>Constraint</u>: Task is manipulated creating a possible 4v1 affording the defenders an incentive to prevent passes going to the striker.

<u>Coaching points:</u>

- Interactions between teammates.
- In the middle third, one player pressing the ball and the other two cutting passing lanes.
- Trying to stay chest onto the attacker in 1v1 situations.
- Decision-making of when to press the ball and when to cut passing lanes.
- Force the attacker wide and backward in 1v1 situations.

3v3 high press: SSG

Session created with SportsSessionPlanner.com

Representative design 6/10   Intentionality 7/10
Affordances 8/10   Repetition without repetition   8/10

<u>Session organization</u>: In a small area, set up a 3v3 with mini goals and no goal-keepers. The aim is to score in the mini goal. The additional rule is that if a team wins the ball back in the opposition half, that is worth one goal, and if they score from winning the ball in that half, that goal is worth three.

<u>Problem</u>: Goals are worth more if a team wins the ball in the opponents' half.

<u>Solution</u>: Press high in the opponents' half to try to win the ball and try to score.

<u>Increase levels of complexity</u>:
• Increase number to a 4v4.
• Increase the area size both depth and width after every goal scored.

<u>Decrease levels of complexity</u>:
• Decrease numbers to a 2v2.
• One player from each team must stay in each half when in possession.

<u>Constraint</u>: Task is manipulated by the rule that if you win the ball back in the opposition half, that is worth one goal, and if you score from it, that is worth three, affording more opportunities to press high.

<u>Coaching points:</u>

- Delaying the opponent.
- Force wide and backward in 1v1 situations.
- Interactions between teammates.
- Decision-making of when to jockey, tackle, or intercept.
- Pressing triggers.

High pressing: Specific

Session created with SportsSessionPlanner.com

Representative design   7/10   Intentionality   7/10
Affordances   8/10   Repetition without repetition   8/10

Session organization: In half a pitch, set up a 6v4 with an additional two on the halfway line. The team of four defenders aim to press the ball; after 6 seconds, two more defenders join to support the press. The team of six attackers aim to play the ball into a mini goal located in either corner on the halfway line. The ball begins with the goalkeeper to maintain realism. The attacking team that start in possession are awarded two goals if they can score in a mini goal within 6 seconds.

Problem: Players begin pressing underloaded in a 6v4 situation knowing more players join later if they can delay the attackers.

Solution: Delay the attackers playing out from the back as much as possible before pressing more aggressively in a 6v6 situation.

Increase levels of complexity:

- If defenders win the ball, they have 6 seconds to shoot.
- Vary the start position, sometimes with a full back or center back instead of the goalkeeper.
- If the defenders win the ball and score, one of their players joins the other team.

Decrease levels of complexity:

- Start with a 3v3 and the goalkeeper.
- Add one more player from each team every 10 seconds with the play still beginning with the goalkeeper.
- If the attackers score, one of their players joins the other team.

Constraint: Task is manipulated by having two extra defenders join after 6 seconds, and the team starting in possession should try to play quickly knowing a goal scored within 6 seconds worth two.

Coaching points:

- Delay the opponent in 1v1 situations.
- Interactions between teammates.
- Force the attackers sideways or backward.
- Decision-making of when to press, tackle, or intercept.
- Pressing triggers.
- Large strides to cover big spaces but short strides when dealing with a 1v1 situation.

High pressing: SSG

Session created with SportsSessionPlanner.com

Representative design 7/10   Intentionality   7/10
Affordances   8/10   Repetition without repetition   7/10

Session organization: In half a pitch, set up a 7v6 with the two corners of the pitch removed. Both teams aim to score in a full-sized goal located centrally; the team with seven is attacking the goal where the corners have been removed. A goal scored within 6 seconds from winning the ball is worth three. Players from each team can be changed around as the problems for each team are different.

Problem: Lack of width at one end of the pitch

Solution: Team of 7 press high up the field

Increase levels of complexity:

- If a team wins the ball in a high press, then a member of the opposition must sit out the next phase of play.
- Winning the ball in a high press is worth one.
- If a goal comes from winning the ball in a high press, it is worth three.
- After a team scores a goal, one of their players joins the other team.

Decrease levels of complexity:

- Every play starts with goalkeeper of the team with six in the form of a goal kick.
- Make one player neutral so they play for the team in possession.

Constraint: Task is manipulated by the pitch dimensions to create different problems for both teams. Team of six must force the opposition wide to and then cut off passing lanes into the center. This incorporates the idea of delaying the opponent before aggressively trying to win possession. Team of seven are encouraged to press high as their opponents do not have much width to use.

Coaching points:

- Force the attackers wide and backward.
- Interactions between teammates.
- Decision-making of when to jockey, tackle, or intercept.
- Staying chest on with attackers in 1v1 situations.
- Pressing triggers.
- Large strides to cover big spaces but short strides when dealing with a 1v1 situation.

References

Andersen, T., & Dörge, H. (2011). The influence of speed of approach and accuracy constraint on the maximal speed of the ball in soccer kicking. *Scandinavian Journal of Medicine & Science In Sports, 21*(1), 79−84. https://doi.org/10.1111/j.1600-0838.2009.01024.x.

Araújo, D., Davids, K., & Hristovski, R. (2006). The ecological dynamics of decision making in sport. *Psychology of Sport and Exercise, 7*(6), 653−676. https://doi.org/10.1016/j.psychsport.2006.07.002.

Araujo, D., Rocha, L., & Davids, K. (2010). The ecological dynamics of decision making in sailing. In I. Renshaw, K. Davids, & G. Savelsbergh (Eds.), *Motor learning in practice: A constraints led approach (pp. 131−143).* London: Routledge.

Araújo, D., Ripoll, H., & Raab, M. (2009). *Perspectives on cognition and action in sport.* Hauppauge, NY: Nova Science.

Barris, S., Farrow, D., & Davids, K. (2013). Do the kinematics of a baulked take-off in springboard diving differ from those of a completed dive. *Journal of Sports Sciences, 31*(3), 305−313. https://doi.org/10.1080/02640414.2012.733018.

Burton, D., & Raedeke, T. (2008). *Sport psychology for coaches.* Champaign: Human Kinetics.

Butler, J. (2005). TGfU pet-agogy: Old dogs, new tricks and puppy school. *Physical Education and Sport Pedagogy, 10*(3), 225−240. https://doi.org/10.1080/17408980500340752.

Chow, J., Davids, K., Button, C., & Renshaw, I. (2016). *Nonlinear pedagogy in skill acquisition* (pp. 25−44).

Daniels, H. (2005). *An introduction to Vygotsky.* London: Routledge.

Davids, K. (2011a). The constraints-based approach to motor learning: Implications for non-linear pedagogy. In I. Renshaw, K. Davids, & G. Savelsbergh (Eds.), *Motor learning in practice: A constraints-led approach* (pp. 3−16). London: Routledge.

Davids, K. (2011b). The constraints based approach for motor learning: Implications for a non-linear pedagogy in sport and physical education. In I. Renshaw, K. Davids, & G. J. P. Savelsbergh (Eds.), *Motor learning in practice: A constraints-led approach* (1st ed.). Routledge.

Davids, K., Button, C., & Bennet, S. (2008). *Dynamics of skill acquisition: A constraints led approach.* Human Kinetics.

Davids, K., Hristovski, R., & Araujo, D. (2015). *Complex systems in sport.* London: Routledge.

Dicks, M., & Chow, J. (2011). A constraints-led approach to coaching association football: The role of perceptual information and the acquisition of co-ordination. In I. Renshaw, K. Davids, & G. Savelsbergh (Eds.), *Motor learning in practice: A constraints led approach* (1st ed., pp. 161−172). London: Routledge.

Gibson, J. (1979). *The ecological approach to visual perception.* Boston: Houghton Miffin.

Hargreaves, A., & Bate, R. (1990). *Skills and strategies for coaching soccer.*

Hill-Haas, S., Dawson, B., Impellizzeri, F., & Coutts, A. (2011). Physiology of small-sided games training in football. *Sports Medicine, 41*(3), 199−220. https://doi.org/10.2165/11539740-000000000-00000.

Hunter, A., Angilletta, M., & Wilson, R. (2018). Behaviors of shooter and goalkeeper interact to determine the outcome of soccer penalties. *Scandinavian Journal of Medicine and Science in Sports, 28*(12), 2751−2759. https://doi.org/10.1111/sms.13276.

Lees, A., & Owens, L. (2011). Early visual cues associated with a directional place kick in soccer. *Sports Biomechanics, 10*(02), 125−134. https://doi.org/10.1080/14763141.2011.569565.

Magius, T., Pill, S., & Elliot, S. (2015). An application of non-linear learning in netball: Game-sense coaching. In *29th Achper international conference* (pp. 235–247). South Australia: Flinders University.

Martindale, R., & Nash, C. (2013). Sport science relevance and application: Perceptions of UK coaches. *Journal of Sports Sciences, 31*(8), 807–819. https://doi.org/10.1080/02640414.2012.754924.

Newell, K. (1986). Constraints on the development of coordination. *Motor Development in Children; Aspects of Coordination and Control*, 341–360.

Newell, K., & Ranganathan, R. (2011). Instructions as constraints in motor skill acquisition. In I. Renshaw, K. Davids, & G. Savelsbergh (Eds.), *Motor learning in practice: A constraints-led approach* (pp. 17–32). London: Routledge.

Pinder, R. (2011). The changing face for practice developing perception: Action skills in cricket. In I. Renshaw, K. Davies, & G. Savelsbrough (Eds.), *Motor learning in practice: A constraints-led approach* (pp. 99–108). London: Routledge.

Pinder, R., Renshaw, I., & Davids, K. (2009). Information–movement coupling in developing cricketers under changing ecological practice constraints. *Human Movement Science, 28*(4), 468–479. https://doi.org/10.1016/j.humov.2009.02.003.

Renshaw, I. (2011). Building the foundations: Skill acquisition in children. In I. Renshaw, K. Davids, & G. Savelsbergh (Eds.), *Motor learning in practice: A constraints led approach* (pp. 33–45). London: Routledge.

Renshaw, I., Araújo, D., Button, C., Chow, J., Davids, K., & Moy, B. (2015). Why the constraints-led approach is not teaching games for understanding: A clarification. *Physical Education and Sport Pedagogy, 21*(5), 459–480. https://doi.org/10.1080/17408989.2015.1095870.

Renshaw, I., Chow, J., Davids, K., & Hammond, J. (2010). A constraints-led perspective to understanding skill acquisition and game play: A basis for integration of motor learning theory and physical education praxis? *Physical Education and Sport Pedagogy, 15*(2), 117–137. https://doi.org/10.1080/17408980902791586.

Renshaw, I., Davids, K., Newcombe, D., & Roberts, W. (2019). *The constraints led approach: Principles for sports coaching and practice design* (1st ed.). London: Routledge.

Sampaio, J., Lago, C., Gonçalves, B., Maçãs, V., & Leite, N. (2014). Effects of pacing, status and unbalance in time motion variables, heart rate and tactical behaviour when playing 5-a-side football small-sided games. *Journal of Science and Medicine in Sport, 17*(2), 229–233. https://doi.org/10.1016/j.jsams.2013.04.005.

Savelsbergh, G., Versloot, O., Masters, R., & van der Kamp, J. (2011). Saving penalties, scoring penalties. In I. Renshaw, K. Davids, & G. Savelsbergh (Eds.), *Motor learning in practice: A constraints-led approach* (pp. 57–68). Lodnon: Routledge.

Stevens, T., De Ruiter, C., Beek, P., & Savelsbergh, G. (2015). Validity and reliability of 6-a-side small-sided game locomotor performance in assessing physical fitness in football players. *Journal of Sports Sciences, 34*(6), 527–534. https://doi.org/10.1080/02640414.2015.1116709.

Tan, T., Chow, J., Duarte, R., & Davids, K. (2017). Manipulating task constraints shapes emergence of herding tendencies in team games performance. *International Journal of Sports Science and Coaching, 12*(5), 595–602. https://doi.org/10.1177/1747954117727661.

Vaeyens, R., Lenoir, M., Williams, A. M., Mazyn, L., & Philippaerts, R. M. (2007). The effects of task constraints on visual search behavior and decision-making skill in youth soccer players. *Journal of Sport and Exercise Psychology, 29*(2), 147–169.

Vuuren-Cassar, G., Lachance, E., & Poulin, C. (2014). Karen. In K. Armour (Ed.), *Pedagogical cases in physical education and youth sport* (pp. 263–276). London: Routledge.

Index

Printed in the United States
By Bookmasters